Lee or Grant:

Who Was the Superior General of the Civil War?

By

David W. Manly

David W. Manly

This book is dedicated to the memory of:

Franklin B. Manly

Great, Great Grandfather
1839-1914
86th Regiment Ohio Infantry and
Co B 161 Regiment of Ohio Infantry

George Washington Barnhart

Great Grandfather
1841-1928
Co E 74th Regiment of Indiana Infantry

And in appreciation of my wife Marcia, and children Sarah, Jordan and Will who have accompanied me on hot and dusty battlefields—and to grandchildren Ellie, Jack and George who may enjoy this in the future.

David W. Manly

Hardcover ISBN: 979-8-9956328-0-1
Paperback ISBN: 979-8-9956328-1-8

Order on-line at www.davidwmanly.com or www.amazon.com

Author inquiries at www.davidwmanly.com or via e-mail to the author at leeorgrant@gmail.com.

David W. Manly

"It is well that war is so terrible, or we should grow too fond of it."

Robert E. Lee to James Longstreet at the Battle of Fredericksburg, 1862

"The art of war is simple enough. Find out where your enemy is. Get at him as soon as you can. Strike him as hard as you can, and keep moving on."

Ulysses S. Grant to a reporter during the Battles of Ft. Henry/Ft. Donelson, 1862

Table of Contents

List of Maps

(Note: some maps are shown horizontal and some vertical to optimize readability in this book format)

Introduction

This book seeks to determine who was the superior general of the U.S. Civil War. There really are only two candidates to consider. Was it Robert E. Lee who was the preeminent general for the Confederate States of America (CSA) or Ulysses S. Grant who rose thru the ranks of the Union Army to the position of General in Chief of Union forces by the end of the war?

Military historians, biographers, journalists, and ordinary citizens have pondered the question of which general deserves the top spot on the podium for the last 160 years.

Soon after the war ended Lee's reputation greatly outshone that of Grant, as many of the post-war writings were by former Confederate generals providing their view of the conflict. Lee was portrayed as a master tactician that only lost because of the superior resources of the North, and failures of his "lieutenants". This was the so-called "Lost Cause" argument that prevailed at the end of the 19th century and early to mid-20th century.

Conversely, Grant was portrayed as a clumsy "butcher" that squandered his troops in bloody battles but won because of the almost unlimited ability to raise new troops and material resources in the more populated north. More recently though, historians have reconsidered Grant. Grant biographies by historians such as Ron Chernow, Ronald White, and Brooks Simpson have pushed back on the Lost Cause argument.

So who is right? The Lost Cause proponents or the more recent historians pushing back on the Lost Cause. As an amateur Civil War aficionado, I have never found a succinct, persuasive book or article that reviews the performance of Lee and Grant in a dispassionate way to isolate who was the superior general in the war. So I have set out to do it myself.

In this book we take a new look at Lee and Grant. This is not a history book with either veiled or outright partisanship in every chapter. Rather, this book tries to look OBJECTIVELY at the facts to determine whether Lee or Grant stands highest on the podium of our Civil War history.

I am not a military person, or a historian. I am just a retired businessman who has had a lifelong interest in the Civil War since discovering that a Great Grandfather on my mother's side and a Great, Great Grandfather on my father's side volunteered to fight in the war. I have DNA connection--like many of you readers.

David W. Manly

There are literally thousands of books, articles, and web posts that seeks to answer the question of which General deserves the top spot. But to me, none of the readings I have seen present the question in a manner I am used to in business. In business we make decisions based on assembling as many facts as possible and then using those facts to make a reasoned decision. As much as possible we try to eliminate emotion or personal bias from the process.

My career was in building consumer products. Early on I worked at Procter & Gamble (P&G) leading the marketing of brands such as Pringle's Potato Chips, Crisco, and Duncan Hines. We were trained to use data to make decisions. What was our market share? What factors drove share growth or declines? Did our advertising work to drive our strategic messaging? Everything we did was data driven with no emotion or bias. Result: P&G competed in 50 product categories and had the leading brand in 40. Not bad. My early training at P&G helped me build later brands such as helping to make the Keurig single cup coffee system the number one coffee maker in North America in just a few short years.

So I have applied the same tenets used in business to answer the age-old question of "who was best" by attempting to develop and analyze DATA to form a conclusion. More about that in the following brief *Methodology* section.

I hope you enjoy what follows. While my ancestors fought for the Union, my efforts here strive to be as impartial as possible. I had no idea where it would turn out until completing the analysis. Of course, anyone can disagree with my ratings. That is the fun of being arm chair Civil War historians, right?

David Manly

Duxbury, MA
April, 2026

Methodology and Sources

The book focuses on the major campaigns and battles Lee and Grant participated in in primary leadership roles. Each battle is analyzed individually to isolate the performance of the general. What you won't find is a detailed recap of the battles as found in most history books. Instead there is a brief summary of each campaign/battle followed by evaluation.

The leadership of Lee and Grant in each of their engagements is evaluated on the following criteria as:

1. **Strategy**—what was the primary objective of the General and his army in a given battle/campaign? Was the strategy sound? Did the General accomplish his objective?

2. **Tactics**—how sound were the tactics the General used to accomplish his strategy? Did they work?

3. **Casualties**—what were the respective casualties (killed/wounded/missing/captured) of the General compared to his opposition? Did the General make optimal use of his forces compared to his opponent?

The General's performance in the above criteria is then ranked on a 5-point scale as follows:

Performance Ranking Methodology

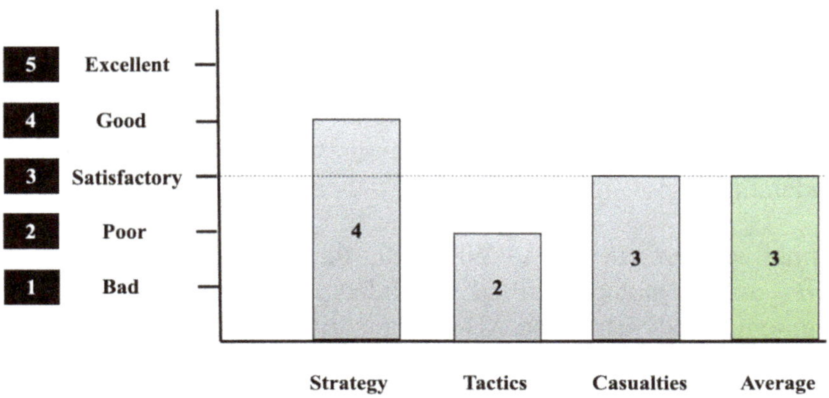

In this manner we can assign "grades" to each General and at the end tally up all the grades to see who the "valedictorian" is between Lee and Grant.

We will analyze each battle/campaign in order of when it occurred during the war. Each section includes the best public source map of the action that I could find. Map sources are noted.

In addition to the ranking metrics above each battle/campaign is further analyzed by reviewing the casualty statistics from the action. In this manner we can see the relative losses each general suffered and those his armies inflicted. This provides a further look at how Lee and Grant utilized their forces compared to their enemy. A table such as the following recounts the data:

Sample Battle Casualty Comparison											
Union						Confederate					
Strength	Killed	Wounded	Captured/ Missing	Total	%	Strength	Killed	Wounded	Captured/ Missing	Total	%

Source Data

In this book I strive to maintain accuracy and consistency in the source data used. There are hundreds of books and articles that provide conflicting data on casualty figures for Civil War battles.

My primary source of data comes from the **American Battlefield Trust** (ABT). The ABT is a leading nonprofit 501(c)(3) organization dedicated to preserving America's historic battlefields from the Revolutionary War, War of 1812, and Civil War. It educates the public about these sites and their significance through various programs.

This book uses the American Battlefield Trust (ABT) Timeline of Major Civil War Battles to list those battles/campaigns where Lee and Grant had primary leadership responsibility.

From the timeline, one can link to Wikipedia that lists data on casualties from the ABT usually stated as killed, wounded, and captured/missing for each engagement. While using the ABT data is just one source, this author judges their data to be fully vetted, and representing the best consensus possible of casualties during the war. I have compared the ABT data to other sources and while at times they can differ, the differences are not enough to change any conclusions about "what happened".

For maps, this book uses royalty free maps and images that are in the public domain. Most of the maps in the book are excellent drawings from Hal Jesperson at www.cwmaps.com.

What you won't find in this book is reems of footnotes and citations linking to an extensive bibliography. I don't pretend to be a scholar—rather an amateur interested in this specific topic. Everything in the text is commonly understood by anyone who reads the major books about the war. The book closes with *Suggested Reading* of books I personally have found most enlightening about the war and Lee and Grant in particular.

Ready to get going? Let's go.

First up though, let me provide a brief biography of each general for those not familiar.

Brief Biographies—Lee and Grant

Robert E. Lee (1807–1870)

Robert Edward Lee was a prominent American soldier, best known as the commander of the Confederate Army of Northern Virginia during the Civil War (1861–1865). He is regarded as one of the most skilled soldiers in U.S. history, though his decision to fight for the Confederacy has made him a controversial figure for many. On the one hand, Lee symbolizes the long and distinguished heritage of the southern states. On the other hand, Lee chose to fight for the Confederate States of America—whose member states cited "maintaining slavery" as a key reason for their secession from the United States.

Early Life and Education

Born on January 19, 1807, at Stratford Hall in Westmoreland County, Virginia, Lee was the son of Henry "Light-Horse Harry" Lee, a Revolutionary War hero and former Virginia governor, and Ann Hill Carter Lee. His family's prestige was tarnished by his father's financial failures and imprisonment for debt, leading the family to move to Alexandria, Virginia. Lee's father left for the West Indies in 1813 and died soon after.

To escape poverty, Lee attended the United States Military Academy at West Point, graduating second in his class in 1829 with no demerits—an exceptional record. He joined the U.S. Army Corps of Engineers.

Robert E. Lee as a young soldier

Pre-Civil War Military Career

Lee's early career involved engineering projects, including coastal fortifications. In 1831, he married Mary Anna Randolph Custis, great-granddaughter of Martha Washington, at Arlington House (her family estate, later Arlington National Cemetery). They had seven children.

Lee gained fame during the Mexican-American War (1846–1848), serving under General Winfield Scott and earning brevet promotions for bravery at battles like Cerro Gordo and Chapultepec. From 1852–1855, he was given the high honor of serving as superintendent of West Point. In 1859, he led troops to suppress John Brown's raid at Harpers Ferry.

Lee inherited enslaved people through his marriage and father-in-law's will. Historical accounts indicate he viewed slavery as a necessary evil, though he privately called it a "moral and political evil." He freed his inherited slaves in 1862 as required by the will.

Civil War

Lee opposed secession but prioritized loyalty to Virginia. Prior to the Civil War most Americans considered themselves more loyal to their states than the Union itself. Offered command of Union forces by Abraham Lincoln in April 1861, he declined and resigned from the U.S. Army after Virginia seceded. Lee simply could not take up a sword against the state of his forebears. He became a Confederate general, initially advising President Jefferson Davis.

In June 1862, Lee assumed command of the Confederate Army of Northern Virginia. His aggressive tactics led to victories in the Seven Days Battles, Second Bull Run, Fredericksburg, and Chancellorsville. He invaded the North twice: in 1862 (halted at Antietam, the war's bloodiest day) and 1863 (defeated at Gettysburg, a turning point).

In 1864 Lee defended his beloved Virginia from Ulysses S. Grant's relentless attacks which came to be known as “The Overland Campaign”. Lee and his forces successfully blunted Grant’s advances but with great, insurmountable losses. A stalemate ensued at Petersburg, Virginia until spring 1865 when Grant renewed his attacks culminating in Lee’s surrender of his army at Appomattox Court House on April 9, 1865.

Lee After Appomattox

Post-War Life and Legacy

Paroled by Grant at Appomattox, Lee became president of Washington College (now Washington and Lee University) in Lexington, Virginia, promoting education and reconciliation.

Lee died on October 12, 1870, from a stroke. Posthumously, he became a central figure in the "Lost Cause" mythology, portraying the Confederacy as a noble defense of "state's rights" despite its association with slavery. Many monuments were erected in his honor, though in recent decades, many have been removed with the rise of "political correctness". Lee remains a complex icon: celebrated for military genius and personal character by some, criticized for treason and ties to slavery by others. Modern historians view him as a product of his time—a dutiful Virginian whose choices were pre-ordained by his upbringing.

Monument to Gen. Robert E. Lee at Gettysburg

Ulysses S. Grant (1822–1885)

Hiram Ulysses Grant (later known as Ulysses S. Grant) was an American military leader and the 18th President of the United States (1869–1877). He commanded Union forces to victory in the American Civil War, accepted Robert E. Lee's surrender, and played a key role in Reconstruction. Despite his military success, his presidency was marred by scandals, though modern historians are starting to rank his presidency higher for his efforts in civil rights and Native American policy.

Early Life and Education

Born on April 27, 1822, in Point Pleasant, Ohio, Grant was the son of Jesse Root Grant, a tanner, and Hannah Simpson Grant. His name was mistakenly recorded as Ulysses S. Grant at West Point due to a correspondence error, and he was given the choice of entering with that name or returning home. He chose to enter—a choice that greatly impacted U.S. history.

Grant graduated from West Point in 1843, ranking 21st in a class of 39. If we were solely to judge Lee and Grant on their metrics at West Point Lee would be the hands down winner! Nicknamed "Sam" by classmates, Grant excelled in horsemanship at the academy.

Grant as a young soldier

Pre-Civil War Military Career

Like Lee, Grant served with distinction in the Mexican-American War (1846–1848) under Generals Zachary Taylor and Winfield Scott, earning brevet promotions for bravery at battles like Molino del Rey and Chapultepec. Interestingly, at Appomattox Grant and Lee reminisced about the Mexican-American War and the fact they met each other at the time.

In the intervening years after the Mexican-American War and before the Civil War Grant fell on hard times and developed a reputation for drinking that would dog him the rest of his life and into posterity.

In 1848 Grant married Julia Dent, the daughter of a Missouri farmer and slave-holder. Initially Julia and their young family were able to accompany Grant on his postings. However, in 1854 Grant was posted to California thousands of miles from his family who remained in St. Louis. During the year he only received one letter from Julia as mail had to be delivered by ship around Cape Horn. Julia was pregnant when he left St. Louis and he had no word of the family or the new child. Grant became lonely and depressed and on occasion abused alcohol.

One day Grant reported under the influence at rollcall and was offered the choice of a court martial or resignation from the Army. Grant resigned and returned penniless to St. Louis. Army veterans heard through their grapevine that Grant had essentially been discharged for being a "drunkard".

Thereafter, whenever it was convenient, Grant's enemies would charge that Grant "was up to his old ways" as a drunk. In reality there are no credible accounts of Grant being inebriated during the Civil War while leading his troops in battle, or afterword as President. After his lonely sojourn to California he was for the most part able to control any desire to drink.

There are credible accounts of two occasions when he "fell off the wagon" during lulls between campaigns when he was inactive but that is all. Many sources claim that when one politician encouraged Lincoln to fire Grant due to his love of drink, Lincoln is reported to have refused stating he could not spare Grant because "he fights". And, according to the anecdote, Lincoln offered to proffer Grant's liquor to other generals if it would have the same impact on them that it did on Grant.

Once out of the army and back to St. Louis, Grant struggled miserably in civilian life. He tried farming, real estate, bill collecting, and selling firewood in St. Louis. Finally, in 1860, Grant moved to Galena, Illinois, working ignominiously as a clerk under his two younger brothers in their family leather business.

Prior to the Civil War Grant's attitude toward slavery is cloudy. In the late 1850s, while living in Missouri and struggling financially as a farmer on his father-in-law's White Haven estate, Grant acquired an enslaved man named William Jones (likely purchased from or given by his father-in-law, Frederick Dent). Both Jones and Grant worked side by side on Grant's farm.

In 1859 Grant personally wrote and filed a manumission document at the St. Louis Courthouse, legally freeing William Jones from slavery forever. This occurred more than two years before the Civil War started in 1861. Notably, Grant could have sold Jones for a significant sum (estimated at $1,000–$1,500, a large amount equivalent to tens of thousands of dollars today) to ease his money troubles, but he chose to emancipate him without compensation instead. So Grant used and profited from slave labor but also freed his slave at a financial loss.

Civil War

At the war's outbreak in 1861 Grant tried to receive a commission in the army, but his past reputation for drink worked against him. He travelled to Cincinnati in an attempt to meet with newly commissioned General McClellan—however McClellan made Grant sit and wait all day and then refused to see him as he knew of Grant's history.

Finally, late in 1861 the State of Illinois needed experienced soldiers to train volunteers and the governor offered Grant a commission. Grant was successful at leading a new regiment and soon found himself leading troops into battle for the first time at the battle Belmont. Promoted rapidly, he captured Fort Donelson in 1862 (earning the nickname "Unconditional Surrender" Grant) and won at Shiloh despite heavy losses.

His masterpiece was the Vicksburg Campaign (1863), capturing the Mississippi River fortress and splitting the Confederacy. President Lincoln appointed him general-in-chief of all the Union Army in 1864.

Grant as General in Chief—1864

Grant's Overland Campaign (1864–1865) wore down Lee's army through attrition, leading to the siege of Petersburg, Virginia and Lee's eventual surrender as a result of the Appomattox Campaign. Grant's generous terms pardoned the Confederate soldiers allowing them to return home with no fear of retribution for treason.

Grant and Lee at Appomattox

Presidency (1869–1877)

Elected as a Republican in 1868 (and reelected in 1872), Grant's administration focused on Reconstruction, enforcing civil rights for freed slaves via the 15th Amendment and the anti-Ku Klux Klan Act. He crushed the first KKK and established the Department of Justice. Foreign policy successes included the Treaty of Washington (1871), settling Alabama claims with Britain. He advocated "Peace Policy" toward Native Americans, though it was inconsistently applied.

Grant’s presidency was plagued by corruption scandals (e.g., Crédit Mobilier, Whiskey Ring), often involving appointees rather than Grant personally, who was honest but naive in trusting friends. Upon a visit to Washington, Robert E. Lee called on Grant at the White House. It was their last meeting.

Grant as President

Post-Presidency and Legacy

After a world tour (1877–1879), Grant lost his fortune in a fraudulent investment firm in 1884. Diagnosed with throat cancer, he raced to complete his memoirs to support his family. *The Personal Memoirs of U.S. Grant* (1885), praised by Mark Twain and modern critics as one of the finest military memoirs, became a bestseller.

Grant died on July 23, 1885, in Mount McGregor, New York. He is buried in Grant's Tomb in New York City, the largest mausoleum in North America.

Grant's Tomb in New York City

Once viewed as a great military leader but poor to mediocre president, recent scholarship (e.g., by Ron Chernow and H.W. Brands) has rehabilitated Grant's reputation, emphasizing his military brilliance, commitment to equality, and personal integrity amid adversity. He remains a symbol of perseverance and Union preservation.

Lee and Grant's Major Campaigns and Battles—In Chronological Order

Grant: *Battle of Belmont* (November, 1861)

Belmont was Grant's initiation as a leader of Union troops in combat. The battle began Grant's ascending learning process from smaller to larger battles/campaigns that led to his eventual leadership of all Union armies.

Synopsis

The **Battle of Belmont**, fought on November 7, 1861, in Mississippi County, Missouri, was a small but bloody engagement along the Mississippi River, opposite a Confederate stronghold at Columbus, Kentucky.

Grant, commanding 3,114 Union troops from Cairo, Illinois, steamed downriver with gunboat support on November 6 to disrupt Confederate operations. He suspected reinforcements might be moving from Columbus to support other Southern forces. The next morning, Grant's force landed near Hunter's Farm, about three miles north of the Confederate camp at Belmont (Camp Johnston), to avoid heavy guns on the Kentucky bluffs.

By 9 a.m., Union forces advanced through wooded terrain and attacked the roughly 5,000 Confederates under Brigadier General Gideon J. Pillow (part of Major General Leonidas Polk's command). The inexperienced Union troops initially succeeded, routing the Southerners from their camp, overrunning the position, and destroying supplies, tents, and equipment they could not carry. Celebrations broke out as soldiers looted and set fires.

However, disorder set in. Polk rushed reinforcements across the river, and Confederate artillery from Columbus shelled the Union position. The reformed Confederates counterattacked fiercely. Grant's men, low on ammunition and facing superior numbers, fell back in confusion toward their river transports. Panic spread; some officers feared encirclement. Grant remained calm, famously declaring they had cut their way in and could cut their way out. He was among the last to board the riverboat, narrowly escaping as bullets flew all around him but narrowly missed.

Casualties were nearly even and heavy for such a small fight: Union losses totaled 607; Confederates suffered 641. Both sides claimed victory. Confederates celebrated repulsing the invaders and holding the field, while Grant reported to superiors a "complete" success for disrupting enemy plans and gaining field experience for his troops.

Strategically minor, Belmont provided Grant crucial lessons in leadership during the "fog of war," troop discipline, and aggressive action. It boosted Northern morale at a quiet time in the Western Theater and foreshadowed his bolder campaigns at Forts Henry and Donelson. The engagement highlighted the war's growing intensity in the Mississippi River region without altering the broader stalemate.

Grant's Performance: Strengths and Challenges

Grant's performance at Belmont was a mixed but formative experience, showcasing his emerging leadership qualities while exposing tactical inexperience. As a 39-year-old West Point graduate who had seen limited action in the Mexican-American War and resigned from the Army in 1854 due to prior mentioned personal struggles, Belmont marked his first test as a general officer. He demonstrated initiative by launching the raid without explicit orders from his superior, Major General John C. Frémont, to prevent Confederate troops from aiding forces in Arkansas or Missouri.

Key Strengths:

1. **Aggressiveness and Willingness to Fight:** Unlike many early-war Union commanders who avoided risk, Grant pressed the attack aggressively. He speculated post-battle that his advance had deterred Confederate detachments from Columbus, a psychological boost that shaped his bolder strategy in future campaigns like Forts Henry and Donelson.

2. **Personal Bravery:** Grant displayed courage by remaining on the front lines and being among the last to board the retreating transports, narrowly escaping capture as Confederate reinforcements arrived. This act impressed observers and contributed to his reputation.

3. **Troop Morale and Praise:** He commended his inexperienced soldiers for behaving like "seasoned veterans" under fire, issuing a

rare congratulatory order the next day. The engagement provided valuable combat experience to troops who would form the core of the future Army of the Tennessee.

Challenges and Criticisms:

1. **Tactical Errors:** Grant failed to fully anticipate reinforcements from Columbus, leading to a chaotic retreat. His troops, after burning the camp, became disorganized in the dense woods, facing a Confederate counterattack that turned the fight inconclusive. Some historians note "questionable performance" in execution, as the Union burned supplies but couldn't hold ground.

2. **Inconclusive Outcome:** Casualties were roughly equal as shown below. Both sides claimed victory. Confederates touted forcing the Union retreat, while Grant argued it prevented larger enemy maneuvers.

Strategic Impact and Legacy

Though tactically a draw, Belmont boosted Grant's confidence and caught the attention of President Abraham Lincoln, who sought aggressive generals after initial Union setbacks like the first Battle of Bull Run in the eastern theater. It established Grant as a fighter who prioritized action over caution, a trait that propelled him to victories at Shiloh, Vicksburg, and eventual command of all Union armies. The battle also highlighted the role of Union gunboats which provided covering fire during the retreat and foreshadowed river warfare's importance in the Western Theater.

Brigadier General Grant—1861

Belmont remains a pivotal "first test" for Grant, transforming him from an obscure district commander into a rising star. The site is preserved at Columbus-Belmont State Park in Kentucky, commemorating the clash that helped break Kentucky's neutrality and set the stage for Union dominance in the West.

Casualties

Casualties at the Battle of Belmont were roughly even on both sides—607 for the Union and 641 for the Confederates. Grant's losses were 19% compared to a slightly lower 13% for Pillow.

Battle of Belmont Casualty Comparison

Union						Confederate					
Strength	Killed	Wounded	Captured/ Missing	Total	%	Strength	Killed	Wounded	Captured/ Missing	Total	%
3,114	120	383	104	**607**	**19%**	5,000	105	419	117	**641**	**13%**

Rating Grant's Performance at Belmont

Grant's strategic goal of disrupting Confederate reinforcements at Columbus was sound. However, the battle was essentially a "draw" generating a Strategy score of 4 out of 5. Grant's Tactics early on were sound, but fell apart since he failed to consider that the rebels could be reinforced—resulting in a Tactics score of 2 or Poor. His troops were lucky to get out alive. Casualties on each side were equivalent generating a score of 3. Overall, Grant's performance at the early battle of Belmont deserves a *Satisfactory* score of 3.0.

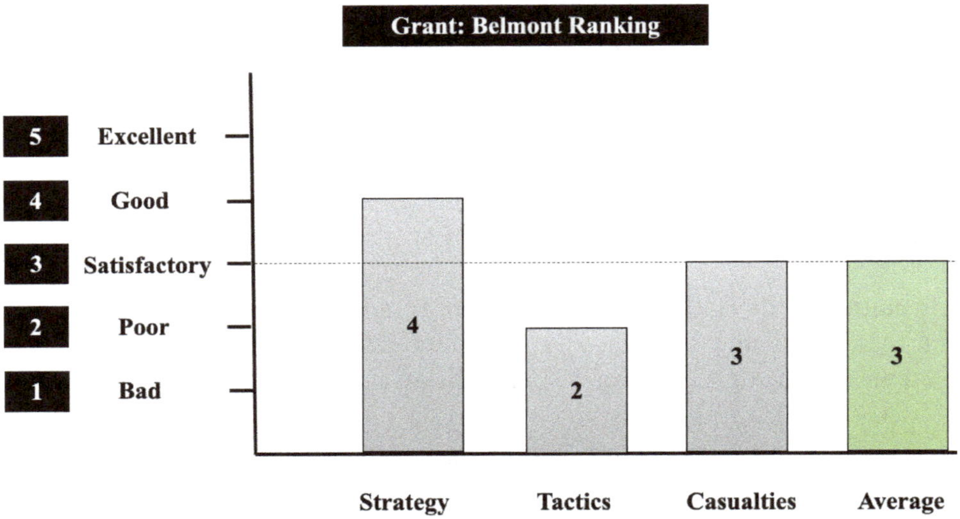
Grant: Belmont Ranking
5 Excellent
4 Good
3 Satisfactory
2 Poor
1 Bad
4
2
3
3
Strategy
Tactics
Casualties
Average

Battle of Belmont

Printed lengthwise for easier reading

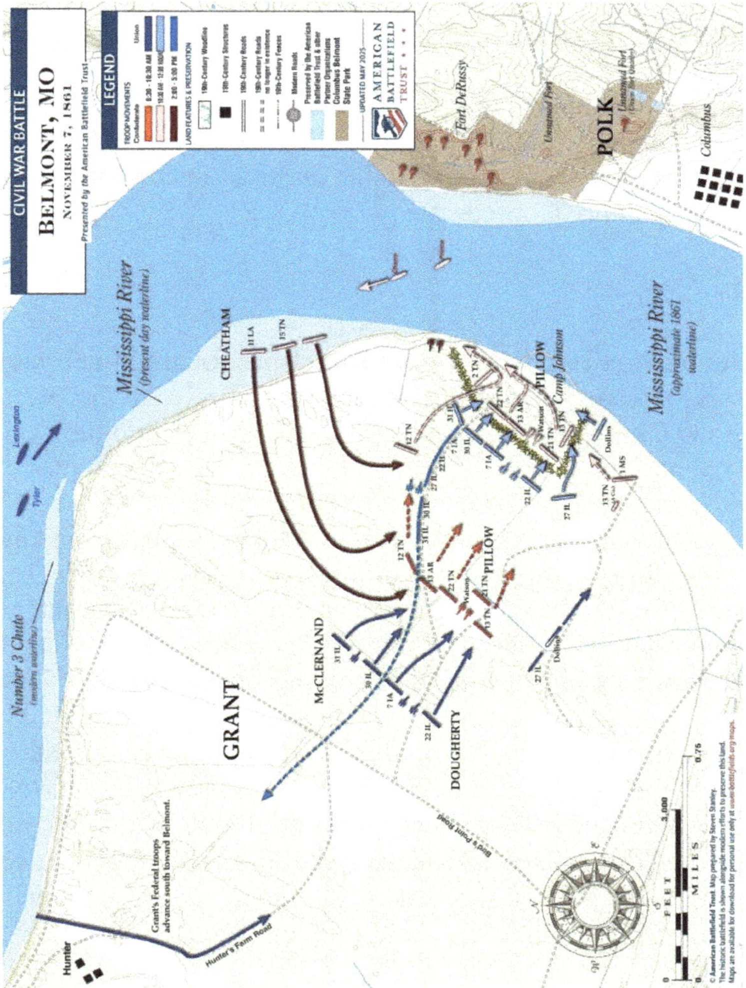

Grant: *Battles of Ft. Henry and Ft. Donelson* (February, 1862)

Ft. Henry and Ft. Donelson launched Grant as a national hero at a time when Union morale was extremely low. Grant's aggressiveness shocked his Confederate opponents leading to the famous "unconditional surrender" of an entire Confederate garrison.

Synopsis

The **Battles of Fort Henry and Fort Donelson**, fought in February 1862 in Tennessee, represented Grant's first major successes in the Western Theater of the Civil War and opened vital river routes into the Confederate heartland.

Fort Henry, an earthen fort on the Tennessee River, guarded a key Confederate defensive line alongside Fort Donelson on the Cumberland River, about 12 miles away. Brigadier General Ulysses S. Grant, commanding roughly 15,000 Union troops, coordinated with Flag Officer Andrew H. Foote's gunboat flotilla (including four ironclads).

The fort was poorly sited and partially flooded by high river waters. On February 6, Foote's gunboats bombarded the position from the river. Confederate commander Brig. Gen. Lloyd Tilghman, with only about 3,200 men, quickly realized resistance was futile as most of his heavy guns were disabled. After a short fight, Tilghman surrendered the fort to the navy before Grant's infantry could fully engage.

Casualties were light: Union losses totaled about 48 (mostly from a boiler explosion on the ironclad USS Essex); Confederates suffered around 79 casualties plus 80–90 captured. Most of the garrison escaped to reinforce Fort Donelson. The victory gave the Union control of the Tennessee River, allowing gunboats to raid deep into Alabama and Mississippi.

Grant then marched his army overland to invest **Fort Donelson**, which sat on higher, drier ground with stronger defenses and a larger garrison that grew to about 16,000 men under Generals John B. Floyd, Gideon J. Pillow, and Simon B. Buckner.

Fighting began on February 13 with Union probes. On February 14, Foote's gunboats attacked the river batteries but were repulsed with heavy damage; Foote himself was wounded. The next day, February 15, the Confederates launched a strong breakout attack that initially succeeded in opening an escape route toward Nashville. However, poor coordination and indecision by Floyd and Pillow led them to order their troops back into the fort, squandering the opportunity.

Grant, who had been absent at the start of the attack, returned and rallied his men for a counteroffensive, tightening the noose. Realizing their position was hopeless, Floyd and Pillow fled with about 2,000 men, leaving Buckner in command. On February 16, Buckner requested surrender terms. Grant famously replied: "No terms except an unconditional and immediate surrender can be accepted." Buckner complied, yielding over 14,000 prisoners, plus artillery, and supplies. Union casualties for Donelson were only 2,691. Combined with Fort Henry, the campaign cost the Confederacy more than 16,000 men.

These twin victories were catastrophic for the South. They forced the evacuation of Columbus, Kentucky, and Nashville, Tennessee (which fell soon after), secured Kentucky for the Union, and opened the Tennessee and Cumberland Rivers as Union invasion corridors. Grant, promoted to major general, earned the nickname "Unconditional Surrender" Grant and emerged as a rising star. The campaign marked the beginning of Union dominance in the Western Theater and boosted Northern morale after early setbacks elsewhere.

Grant's Performance at Fort Henry

Grant's assault on Fort Henry demonstrated effective joint Army-Navy operations, a skill he honed after earlier missteps. Coordinating with Flag Officer Andrew H. Foote's gunboats, Grant proposed the expedition to his superior, Major General Henry Halleck, with Foote's endorsement securing approval.

Key strengths in Grant's performance included:

1. **Strategic Initiative**: Grant exploited the rivers' vulnerabilities using armored gunboats, pushing an offensive that overwhelmed the poorly sited fort, which was prone to flooding.

2. **Coordination with Navy**: Learning from the Battle of Belmont, where poor communication hindered support, Grant emphasized timely consultation. He commended the Navy's "gallant" role, fostering mutual respect that ensured synchronized actions.

3. **Outcome and Impact**: The quick victory (captured in hours) opened the Tennessee River for Union advances, boosted Northern morale, and set the stage for Donelson. Casualties were minimal, with the Navy bearing the brunt of the fighting.

Criticisms were minor; Grant's ground forces played a secondary role due to delays, but this highlighted the Navy's dominance rather than any leadership failing.

Grant's Performance at Fort Donelson

After Fort Henry, Grant moved his forces immediately to threaten nearby Fort Donelson—a much larger garrison on the Cumberland River. Fort Donelson presented greater challenges: stronger fortifications on bluffs, 17 heavy guns, and extensive landward defenses spanning 2.5 miles.

Grant's strategy involved encircling the fort, using naval support to pin the Confederates against the Cumberland River, and launching infantry assaults. Challenges included:

- A harsh winter storm that caused supply shortages and exposure.
- An ineffective naval attack on February 14, where Foote was wounded, and ironclads withdrew after heavy damage.
- Uncoordinated Union infantry probes on February 13.
- A Confederate breakout attempt on February 15, which nearly succeeded until repelled by Brig. Gen. Lew Wallace's division and Grant's decisive counterattack.

Key decisions showcased Grant's adaptability:

1. **Counterattack and Initiative**: Upon returning from a meeting with Foote and finding his lines broken, Grant ordered an immediate

counteroffensive, authorizing attacks that regained lost ground. He requested naval distraction fire, which Commander Benjamin M. Dove's gunboats provided at long range, aiding the Union push.

2. **Unconditional Surrender Demand**: On February 16, after rejecting armistice proposals, Grant famously replied to Buckner (a West Point classmate): "*No terms except an unconditional and immediate surrender can be accepted. I propose to move immediately upon your works.*" This blunt stance, influenced by subordinates like Brig. Gen. Charles F. Smith and echoing Foote's earlier demand at Henry, forced the first full Confederate army capitulation of the Civil War. Grant softened post-surrender with humanitarian gestures, such as sharing rations, allowing officer side arms, and providing medical care.

The battle secured the Cumberland River leading to Nashville's fall on February 25. Now Grant had a nationally recognized nickname: "Unconditional Surrender Grant—a play on his initials U.S. Grant.

Union Gunboats Attacking Ft. Donaldson

Overall Analysis of Grant's Leadership

Grant's performance across both battles was marked by aggressive initiative, adaptability, and a willingness to learn from inter-Service friction. His evolution in joint operations—moving from Belmont's communication lapses to respectful collaboration with Foote—proved crucial, as Navy firepower and transport complemented Army maneuvers, capturing both forts in just 10 days. He recognized the "blades of shears" synergy between branches: united, invincible; separated, useless.

Praises abound: These victories shattered Confederate defenses in the West, shifted Kentucky and Tennessee to Union control, opened invasion routes (leading to Shiloh and Vicksburg), and prompted Grant's promotion to major general of volunteers. Historians credit his decisiveness for exploiting Confederate disarray and river advantages, arguing the South started "losing the war" here due to irreplaceable losses.

Criticisms are limited in these accounts: Some note over-reliance on the Navy at Henry, initial lack of coordination at Donelson, and post-surrender leniency, but these are overshadowed by the strategic triumphs. Overall, the campaign established Grant as a tenacious commander who rejected gentlemanly protocols for total victory, setting precedents for his later successes.

Casualties

In the combined battles the Confederates suffered over 16,000 casualties compared to less than 3,000 for the Union—a ratio of greater than 4 to 1. Most Confederate losses were “captured”.

Battle of Ft. Henry/Donelson Casualty Comparison

	Union						Confederate					
	Strength	Killed	Wounded	Captured/ Missing	Total	%	Strength	Killed	Wounded	Captured/ Missing	Total	%
Ft. Henry	15,000	48	NA	NA	48	0%	3,200	79	NA	90	169	5%
Ft. Donelson	24,531	507	1,976	208	2,691	11%	16,171	327	1,127	14,417	16,171	100%
Total		**555**	**NA**	**NA**	**2,739**	**11%**		**406**	**NA**	**14,507**	**16,340**	**95%**

Rating Grant’s Performance at Ft. Henry and Ft. Donaldson

Grant’s performance in this early campaign ranks a 5 or Excellent in each ranking category for an *Excellent* score of 5.0 overall. Strategically he accomplished his objective of opening the Tennessee and Cumberland Rivers to the Union and pressing the Confederates out of northwest Tennessee all the way to Nashville. Tactically Grant outfoxed his rebel adversaries capturing an entire army. Grant’s casualties were light at under 3,000 compared to the 14,000 suffered by the Confederacy. A truly lopsided campaign in the Union favor. Grant is on his way.

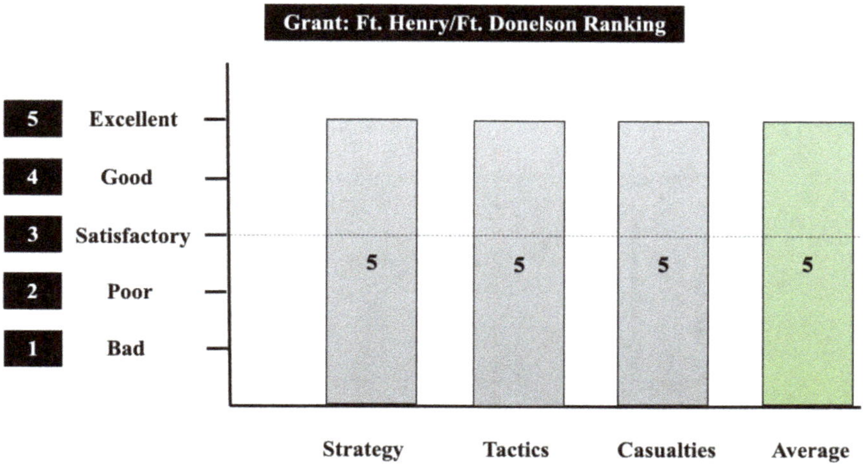
Grant: Ft. Henry/Ft. Donelson Ranking
5
Excellent
4
Good
3
Satisfactory
2
Poor
1
Bad
5
5
5
5
Strategy
Tactics
Casualties
Average

David W. Manly

Battles of Ft. Henry and Ft. Donelson

Map by Hal Jespersen, www.cwmaps.com

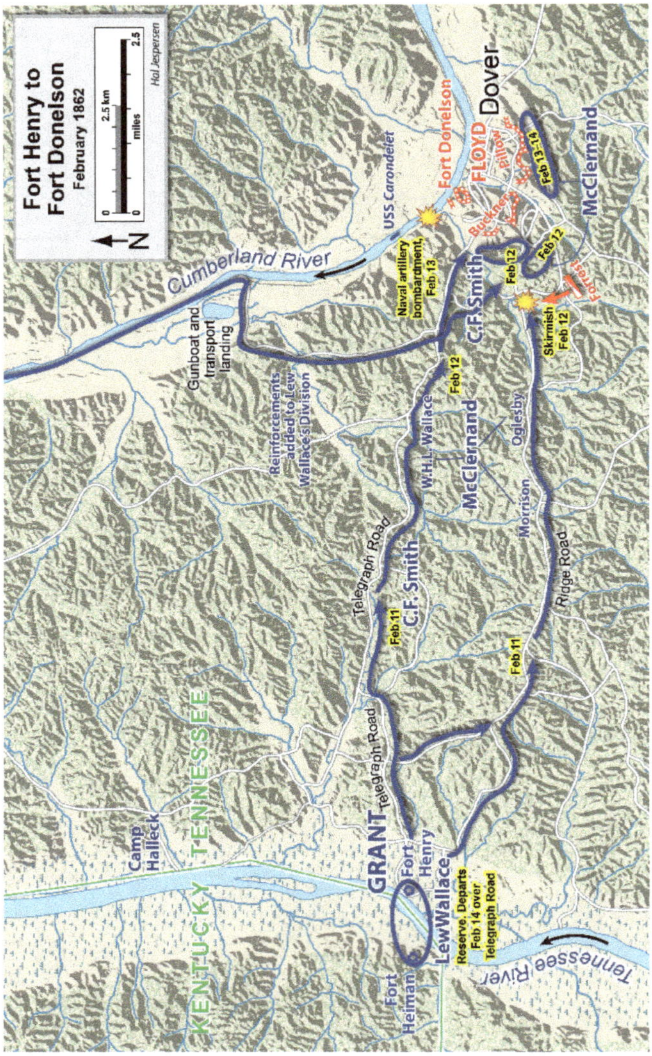

Grant: *Battle of Shiloh* (April, 1862)

The battle of Shiloh was a bloody two-day event that was nearly a disaster for both Grant and the Union cause in the West. Only Grant's steely resolve and tenacity secured victory from the jaws of defeat. The near disaster temporarily halted Grant's meteoric rise in reputation that was only redeemed in future engagements.

Synopsis

The **Battle of Shiloh** (also known as the Battle of Pittsburg Landing), fought on April 6–7, 1862, in Hardin County, Tennessee, was one of the bloodiest and most shocking engagements of the American Civil War up to that point. It marked a major early clash in the Western Theater as Union forces under Maj. Gen. Ulysses S. Grant sought to advance on the vital Confederate rail junction at Corinth, Mississippi.

Following victories at Forts Henry and Donelson, Grant positioned his Army of the Tennessee (about 45,000 men present for duty) at Pittsburg Landing on the west bank of the Tennessee River, near Shiloh Church. He was awaiting reinforcements from Maj. Gen. Don Carlos Buell's Army of the Ohio. Confederate Gen. Albert Sidney Johnston, commanding the Army of the Mississippi (roughly 45,000 men), launched a surprise attack on April 6 attempting to destroy Grant's army before Buell arrived. Heavy rains had delayed the Confederate march, but at dawn they stormed out of the woods, overrunning forward Union camps.

The first day was chaotic and ferocious. Union troops rallied and fought desperate defensive stands at locations like the Peach Orchard, Water Oaks Pond, and the "Hornets' Nest"—a dense thicket where Brig. Gens. Benjamin Prentiss and W.H.L. Wallace held out for hours against repeated assaults. Despite achieving surprise, Confederate forces became disorganized amid the tangled terrain. In the afternoon, Johnston was mortally wounded while leading an attack; command passed to Gen. P.G.T. Beauregard. By nightfall, Grant's battered lines had fallen back to a strong defensive position near the river landing, protected by massed artillery, gunboats, and ravines. Heavy fighting ended in darkness.

Overnight, Buell's reinforcements (about 20,000 men) arrived, giving the combined Union force a significant numerical advantage. On April 7, Grant launched a counteroffensive. The reinforced Federals steadily pushed the exhausted Confederates back across the field in heavy fighting. Beauregard, with his army reduced to roughly 34,000 effectives, skillfully conducted a fighting withdrawal toward Corinth. The Union did not pursue aggressively.

Casualties were staggering: Union losses totaled over 14,047: Confederate losses were over 10,000. The combined total of nearly 25,000 casualties exceeded all previous American wars combined and shocked both North and South, ending illusions of a short conflict.

Though tactically a Union victory, Shiloh was strategically inconclusive in the short term. It allowed the Union to maintain its foothold and later capture Corinth, opening the path toward Vicksburg and eventual Union control of the Mississippi River. The battle intensified Grant's resolve—he later wrote that he gave up any idea of saving the Union except by complete conquest—and established him as a tenacious commander despite criticism for being surprised. For the Confederacy, the death of Johnston was a severe blow, and the heavy losses committed the South to a costly war of attrition in the West.

Union Encampments at Shiloh

The Battle

Day 1: Surprise and Near-Defeat (April 6)

- **The Attack**: At dawn, Johnston's forces struck unexpectedly, overrunning Union camps around Shiloh Church. Grant's army, caught unprepared without strong pickets or cavalry screens, fell back in disarray. Divisions under generals like William T. Sherman and Benjamin Prentiss fought fiercely at key spots (e.g., the "Hornet's Nest"), buying time.

- **Grant's Response**: Grant was having breakfast with his staff when the distant roar of cannon indicated a battle had started. Racing downriver in a steamer and then to the field on horseback, Grant arrived mid-morning and rallied his men with calm resolve. He organized defenses, coordinated counterattacks, and used Union gunboats *Tylor* and *Lexington* to bombard Confederate flanks, halting their advance toward Pittsburg Landing. By evening, Confederates had pushed Union lines back nearly to the river, but Johnston's death from a leg wound (initially mistaken for a minor injury) disrupted their momentum.

- **Criticisms**: Grant was surprised, a point he later conceded in his *Personal Memoirs*, admitting he did not expect the Confederates to go on the offensive. His divided forces (one division under Lew Wallace was delayed miles away) and lack of entrenchments amplified the chaos. Northern newspapers were very critical of Grant for having been surprised and lambasted him for the first time as a "butcher," fueling calls for his removal.

Day 2: Counteroffensive and Victory (April 7)

- **Reinforcements Arrive**: Buell's roughly 20,000 troops crossed the Tennessee overnight, bolstering Grant’s army to about 68,000. With fresh numbers, Grant launched a dawn counterattack.

- **Grant's Leadership**: He directed artillery and infantry advances personally, reclaiming lost ground by afternoon. Brigades under generals like Lovell Rousseau swept forward, driving Confederates back to their Corinth lines. Grant resisted pursuing aggressively into muddy terrain, prioritizing consolidation—a prudent choice given his army's exhaustion.

- **Outcome**: At the end of the battle the Union occupied the field and the Confederate army under Beauregard was in pell-mell retreat back to Corinth. A tactical Union victory, securing Tennessee and paving the way for Corinth's eventual fall. Short-term Grant's reputation hit a speed bump, but long term it boosted Grant's reputation given his failure to yield on the first day, and dogged leadership of the counter attack on Day 2 that secured victory.

Union withdrawal at Shiloh

Assessment of Grant's Performance

Grant's Shiloh tenure is often seen as a mixed but formative test of leadership:

Strengths:

1. **Resilience Under Pressure**: Despite the surprise, Grant never panicked, inspiring troops with his presence (famously replying to a subordinate's dire report: "Yes. Lick 'em tomorrow, though"). His offensive mindset—prioritizing ammunition resupply and gunboat support—turned defense into victory.

2. **Learning Curve**: Shiloh ended Grant's (and the North's) naivety about the war's length. He later wrote in his *Personal Memoirs*

Shiloh convinced him "saving the Union [required] complete conquest." This hardened his resolve, evident in later campaigns like Vicksburg.

3. **Strategic Gain**: Holding the field preserved Union momentum in the West, preventing a Confederate resurgence.

Weaknesses:

1. **Tactical Oversights**: Failure to fortify or screen adequately invited disaster, a recurring issue (e.g., later at the Wilderness). Buell's arrival was crucial; without it, defeat loomed.

2. **Public Backlash**: Initial perceptions of incompetence nearly ended his career, though Lincoln's defense ("I can't spare this man; he fights") saved him.

Historians like Bruce Catton and H.W. Brands praise Shiloh as where Grant "bent, then rebounded," proving his mettle as a commander who thrived in adversity. It wasn't flawless, but it was victorious—and transformative.

Casualties

Casualties on both sides were almost 25,000 making Shiloh the bloodiest battle of the Civil War at that time. While Union losses of 20% were slightly lower than the 24% suffered by the Confederates, Grant losses were about 3,000 more in total. It was a bloody battle all around.

Battle of Shiloh Casualty Comparison

Union						Confederate					
Strength	Killed	Wounded	Captured/ Missing	Total	%	Strength	Killed	Wounded	Captured/ Missing	Total	%
68,812	1,754	8,408	2,885	**13,047**	**20%**	44,699	1,728	8,012	959	**10,699**	**24%**

Rating Grant's Performance at Shiloh

As noted, Grant's performance at Shiloh was mixed. Strategically his objective was to disrupt Confederate transportation and logistics in northern Mississippi by capturing the Corinth rail center. While he eventually achieved his objective, he was forced to fight a terrible battle on terms dictated by his adversary as witnessed by the surprise the army incurred. Strategy score: 4 out of 5.

Tactically Grant rates a satisfactory score of 3 given the surprise and lack of defensive preparations—offset by his aggressive attacks on Day 2 that carried the field in a solid victory. Grant's casualties were higher than the Confederates' due to the first day battle that rolled up his troops—resulting in a score of 2 also.

Net result: Grant at Shiloh—earns a Satisfactory score of 3.0. He turned a looming, catastrophic defeat into a major victory.

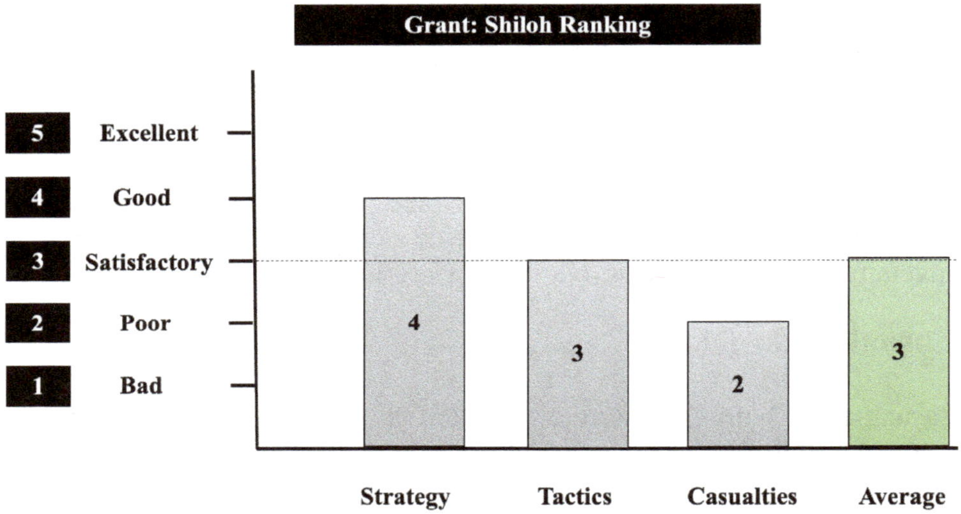

Battle of Shiloh: Day 1

Maps by Hal Jespersen, www.cwmaps.com

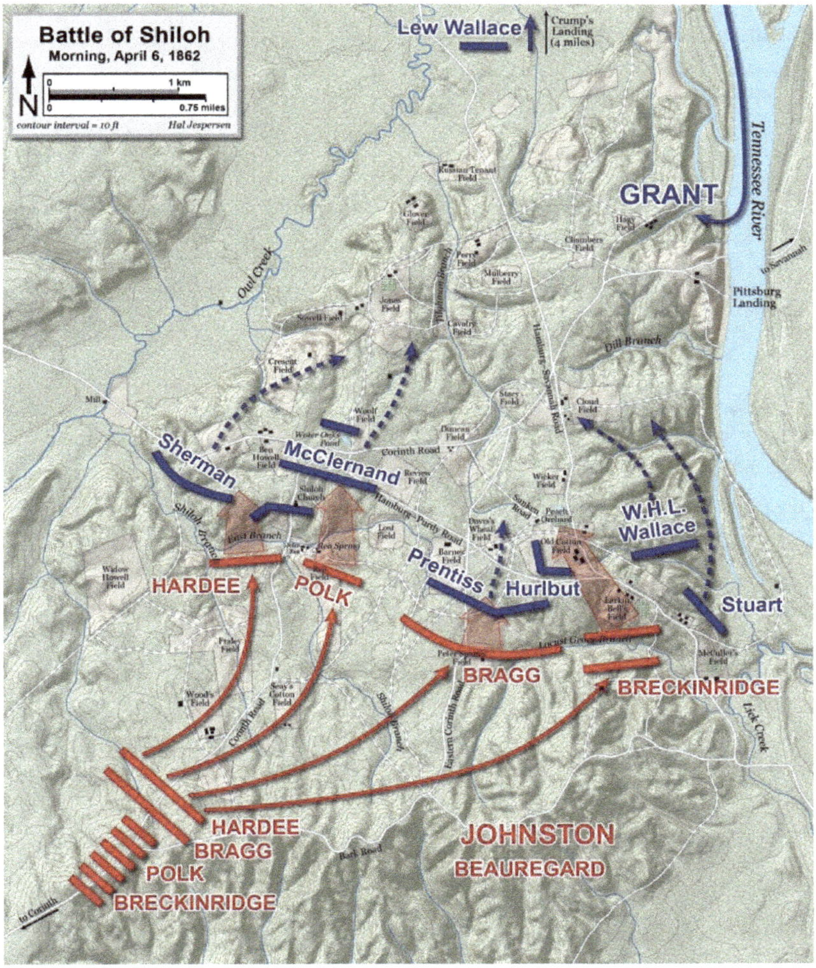

Battle of Shiloh: Day 2

Maps by Hal Jespersen, www.cwmaps.com

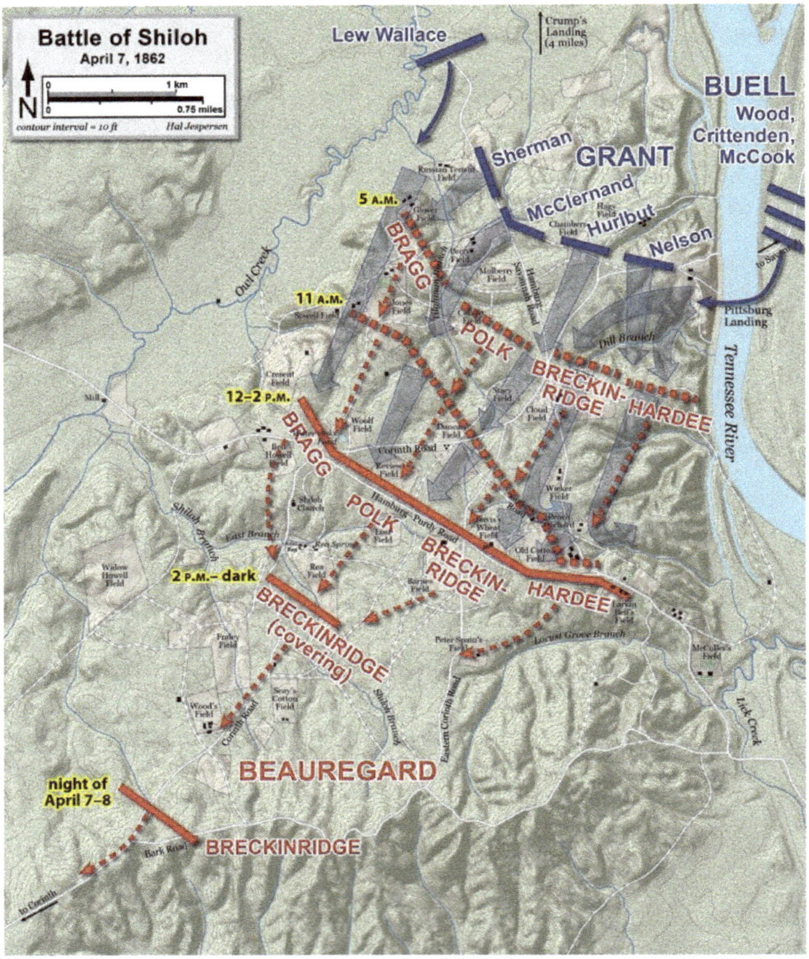

Lee: *The Seven Days Battles* (June/July 1862)

The Seven Days Battles established Robert E. Lee as the preeminent general of the Confederacy. It was the first of many battles where Lee confronted a larger Union force with his smaller army yet was victorious due to his brilliance at tactical improvisation.

Synopsis

The **Seven Days Battles** (June 25–July 1, 1862), the climax of Maj. Gen. George B. McClellan's Peninsula Campaign, were a series of six major engagements east of Richmond, Virginia. Confederate Gen. Robert E. Lee, who had recently assumed command of the Army of Northern Virginia, launched a bold offensive to drive back McClellan's much larger Army of the Potomac (roughly 100,000 men) and save the Confederate capital.

The fighting opened on June 25 with the minor **Battle of Oak Grove**, a limited Union probe. Lee seized the initiative on June 26 at **Mechanicsville** (Beaver Dam Creek), attacking the isolated Union V Corps under Fitz John Porter north of the Chickahominy River. Poor coordination, especially the late arrival of "Stonewall" Jackson's forces, resulted in a Confederate repulse with heavy losses, but it forced McClellan to begin shifting his army.

On June 27, Lee renewed the assault at **Gaines's Mill**, the bloodiest and only clear tactical Confederate victory of the week. After fierce fighting and repeated charges, including the war's largest single Confederate assault (over 50,000 men), Porter's lines broke, compelling McClellan to abandon his supply base on the York River and retreat southeast toward the James River. Minor actions occurred at Garnett's and Golding's Farm on June 27–28.

On June 29, Confederates struck the Union rear guard at **Savage's Station**. The next day, June 30, Lee attempted to intercept the retreating Federals at **Glendale** (Frayser's Farm/White Oak Swamp). Fierce fighting nearly cut the Union army in two, but again poor coordination and Jackson's delays allowed McClellan's forces to escape to a strong defensive position atop **Malvern Hill** near the James River, supported by gunboats.

On July 1, Lee launched repeated frontal assaults against Malvern Hill's commanding heights and massed Union artillery. The attacks were repulsed with terrible Confederate losses in what many called "not war but murder." McClellan, however, chose not to counterattack and continued his withdrawal to the safety of Harrison's Landing on the James River.

Casualties were staggering—the bloodiest week in American history up to that time. Union losses totaled over15,000; Confederate losses were higher at roughly 20,000.

Strategically, Lee achieved a major victory. He lifted the immediate threat to Richmond, ended the Peninsula Campaign, and boosted Southern morale while dealing a severe blow to Northern confidence. Though Lee failed to destroy McClellan's army due to missed opportunities and coordination failures, the campaign established Lee as an aggressive commander and prolonged the war in the East by nearly two years. McClellan's cautious generalship drew sharp criticism from President Lincoln.

George McClellan and Robert E. Lee (Source: Wikipedia)

Key Battles and Lee's Role

Lee's strategy aimed to strike isolated Union corps north of the Chickahominy while pinning down the rest south of the river, forcing McClellan into a general retreat. Coordination faltered due to subordinates like Jackson moving slowly and communication breakdowns, but Lee's persistence paid off. Below is a summary of the battles and Lee's performance:

Lee's Actions: The Seven Days Battles

Dates	Battle	Lee's Key Actions	Outcome	Assessment
25-Jun	**Oak Grove**	Repulsed a minor Union probe south of the Chickahominy: Lee prepared for a flanking attack on the Union right	**Inconclusive**	Set the stage, of minimal impact but showed Lee's initiative
26-Jun	**Beaver Dam Creek (Mechanicsville)**	Directed A.P. Hill and D.H. Hill to assault Union positions; attack failed due to poor terrain	**Union tactical win**	Costly Confederate repulse, but McLellan overacted by withdrawing and ceding initiative to Lee
27-Jun	**Gaine's Mill**	Coordinated a massive assault (50,000+ troops) across the Chickahominy; Jackson delayed but late breakthrough by Longstreet and Hill forced Union retreat	**Confederate Victory**	Lee's first field win; largest battle of the campaign; securing the offensive's momentum
28-Jun	**Garnett's and Golding's Farms**	Magruder (under Lee) demonstrated against Union lines south of the river to feign strength and cover the main pursuit	**Inconclusive**	A diversionary success; it kept McLellan off-balance
29-Jun	**Savage's Station and Allen's Farm**	Pursued retreating Union forces; ordered attacks to harass the withdrawal but held back from full commitment	**Inconclusive**	Slowed Union retreat; Lee adapted to a fluid situation despite lacking intelligence about Union positions
30-Jun	**Glendale (Frazier's Farm)**	Attempted to cut Union retreat route; fragmented attacks due to poor synchronization (Jackson was bogged down in a swamp)	**Inconclusive**	Near miss; heavy losses but prevented Union from consolidating forces
1-Jul	**Malvern Hill**	Launched uncoordinated frontal assaults on fortified Union hilltop; ignored subordinates' warnings about enemy artillery	**Union tactical win**	Lee's worst tactical error; "bloody repulse" but McLellan fled to Harrison's Landing totally abandoning the Union offensive

Confederate Attack on Malvern Hill

Evaluation of Lee's Performance

Strengths:

1. **Strategic Brilliance:** Lee seized the initiative, turning a defensive crisis into an offensive gamble. Despite McClellan's numerical edge, Lee forced a 25-mile Union retreat down the Peninsula, ending the Peninsula Campaign and buying time for Richmond. The outcome set back Union efforts by nearly a year and boosted Confederate morale—his troops dubbed him "Marse Robert" for the victory.

2. **Aggressiveness and Improvisation:** Unlike the cautious Johnston, Lee combined earthworks for defense with rapid maneuvers, exploiting McClellan's paranoia (he believed the Union was outnumbered 2:1). The Gaines' Mill breakthrough exemplified Lee's offensive demeanor in adapting to delays.

3. **Morale and Legacy:** The campaign inspired Lee's army and public confidence, paving the way for later Lee's aggressiveness in future battles like Second Manassas and Antietam. It previewed his high-risk style: dividing forces, accepting casualties for psychological impact.

Weaknesses:

1. **Tactical Flaws:** Poor coordination plagued the offensive—Jackson's sluggishness and Magruder's feints wasted time; Malvern Hill's assaults were "horrific" and unnecessary, with 5,650 casualties in minutes for no gain. Lee's small staff and hands-off approach led to "clumsy" execution.

2. **High Cost:** Confederate losses exceeded the Union's, straining manpower in a resource-poor South. Critics argue Lee's over-aggressiveness foreshadowed issues at Gettysburg.

3. **Missed Opportunities:** Lee couldn't destroy McClellan's army, partly due to poor performance by subordinates and McClellan's timely retreat to the James River.

Overall, Lee's performance was a resounding strategic success that redefined him as the Confederacy's premier general, though tactically uneven. As he later wrote in his official report, the exhausted troops needed rest after seven days of "marching and fighting almost incessantly." The victory ensured the war's prolongation, altering its course.

Despite tactical shortcomings, Lee's leadership in the Seven Days Battles secured a strategic triumph, routing the largest Union army to date and extending the Confederacy's viability by nearly a year. It established him as a formidable commander, shifting the war's dynamics and sustaining Southern hopes amid Western setbacks. However, the campaign's high costs and execution flaws highlighted Lee's limitations in operational details, influencing his later strategies. In broader terms, it underscored how leadership can enable outnumbered forces to achieve outsized results through motivation and risk-taking, though at a steep price.

Casualties

Casualties for the Seven Days Campaign were roughly similar for both armies—at 10% for the Union and 9% for the Confederates. However, Lee lost about 5,000 more men in total than McClellan—losses much harder for the Confederates to replace.

Seven Days Campaign Casualty Comparison

	Union						Confederate					
	Strength	Killed	Wounded	Captured/ Missing	Total	%	Strength	Killed	Wounded	Captured/ Missing	Total	%
Oak Grove	3,000	68	503	55	626	21%	5,000	66	362	13	441	9%
Beaver Dam Creek	15,631	49	207	105	361	2%	16,356	NA	NA	NA	1,484	9%
Gaine's Mill	34,213	1,005	3,107	3,236	7,348	21%	57,018	1,483	6,402	108	7,993	14%
Garnett's/ Golding Farm	3,000	NA	NA	NA	189	6%	5,000	NA	NA	NA	438	9%
Savage's Station	26,600	NA	NA	NA	1,038	4%	14,000	NA	NA	NA	473	3%
Glendale/White Oak Swamp	40,000	297	1,696	1,804	3,797	9%	45,000	638	2,814	221	3,673	8%
Malvern Hill	54,000	324	1,875	818	3,007	6%	55,000	869	4,241	540	5,650	10%
Total		**1,733**	**7,388**	**6,018**	**15,139**	**10%**		**3,056**	**13,819**	**882**	**20,152**	**9%**

Rating Lee's Performance in the Seven Days Campaign

Strategically, Lee deserves an Excellent score of 5. He accomplished his objective of protecting the Confederacy's capital from being captured by the Union which might have ended the war in one fell stroke. Tactically, Lee generally outfought McClellan in the campaign resulting in the Union Army fully abandoning their offensive. Tactical Sore: 5 or Excellent. Lee's casualties were slightly less percentage wise than the Union but total losses were higher for a score of Average or 3. Total rating: *Good* at 4.33.

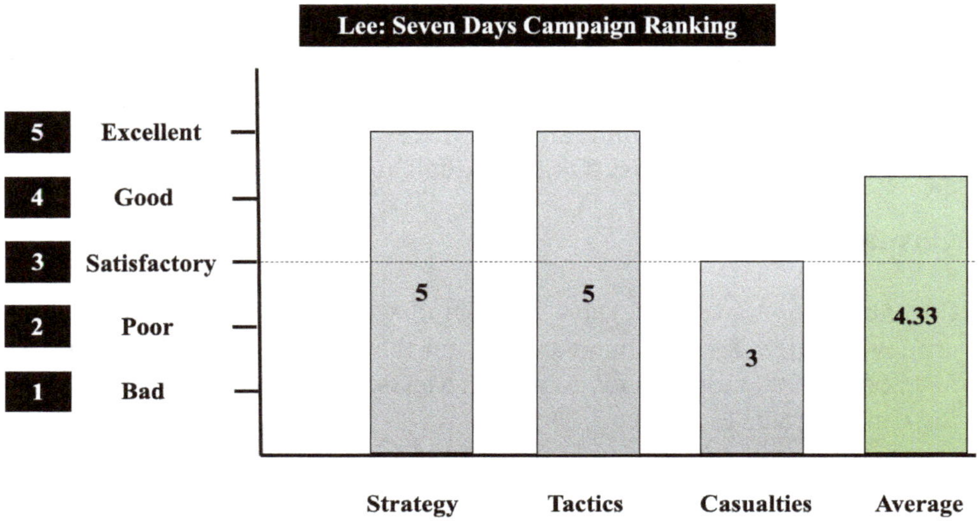

David W. Manly

The Seven Days Battles

Map by Hal Jespersen, www.cwmaps.com

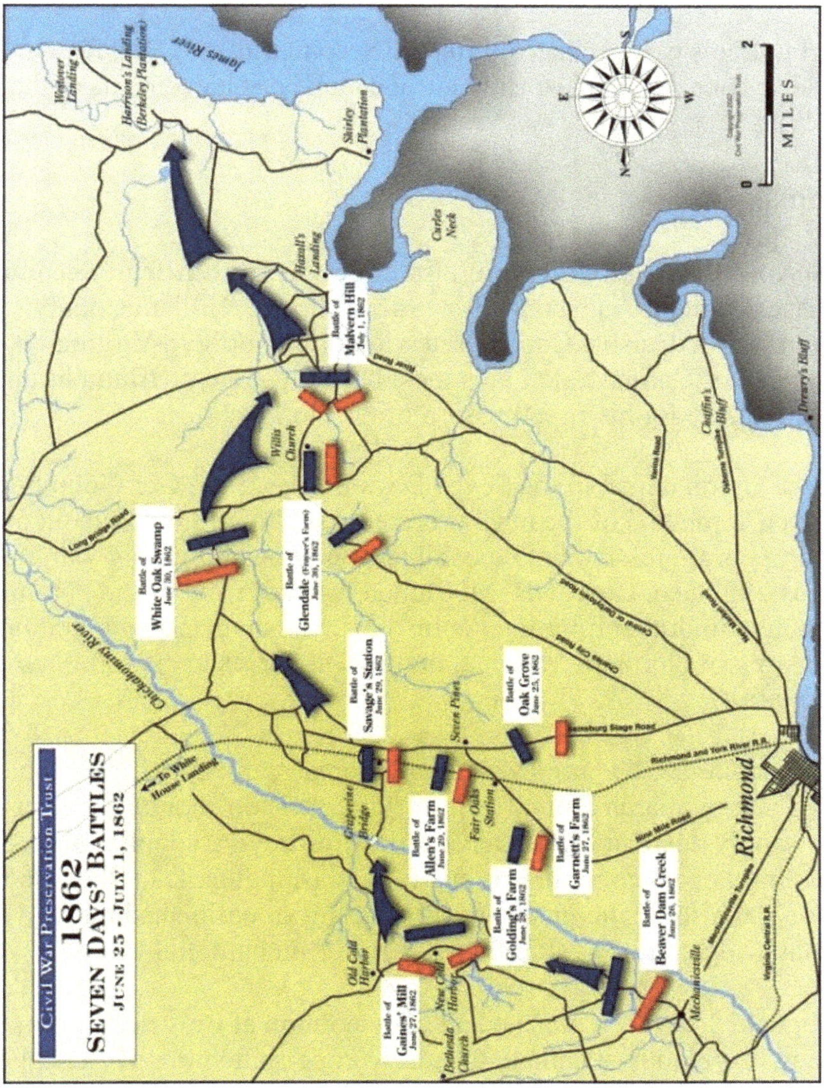

Lee: *Second Battle of Bull Run* (August, 1862)

Robert E. Lee's performance during the Second Battle of Bull Run is widely regarded as one of his finest strategic and tactical achievements of the entire Civil War.

Synopsis

The **Second Battle of Bull Run** (also known as the **Battle of Second Manassas**), fought August 28–30, 1862, in Prince William County, Virginia, was a decisive Confederate victory in Northern Virginia. It marked one of Confederate Lee's most brilliant strategic triumphs and cleared the way for his first invasion of the North.

After the Union defeat in the Seven Days Battles, Maj. Gen. John Pope was given command of the newly formed Army of Virginia (about 50,000 men). Pope's aggressive but boastful style contrasted sharply with the cautious Maj. Gen. George B. McClellan, whose Army of the Potomac was being withdrawn from the Peninsula. Lee, commanding 62,000 men in the Army of Northern Virginia, divided his forces to exploit Pope's vulnerabilities.

On August 25–26, Lee sent Maj. Gen. Thomas "Stonewall" Jackson on a daring flanking march around Pope's right. Jackson captured the massive Union supply depot at Manassas Junction, destroying supplies and threatening Pope's communications with Washington, D.C. Jackson then withdrew to a strong defensive position along an unfinished railroad cut on Stony Ridge, northwest of the First Bull Run battlefield.

On August 28, Jackson attacked a Union column at Brawner's Farm, resulting in a bloody stalemate that drew Pope's attention. Pope, believing he had trapped Jackson, concentrated his forces against him on August 29 and launched repeated assaults against the railroad embankment. Jackson's men repulsed the attacks in fierce fighting, though both sides suffered heavy losses.

Late on August 29, Maj. Gen. James Longstreet's wing of Lee's army arrived after forcing passage through Thoroughfare Gap and took position on Jackson's right. Pope remained unaware of Longstreet's presence. On August 30, believing Jackson was retreating, Pope ordered a general advance. When Maj. Gen. Fitz John Porter's V Corps attacked, Longstreet unleashed a massive counterattack involving nearly 28,000 men—one of the largest simultaneous infantry assaults of the war. The Confederate onslaught crushed the Union left flank and rolled up Pope's line.

Pope's army retreated in disorder toward Centreville and eventually back to the defenses of Washington, D.C. A determined Union rearguard action at Chantilly (September 1) prevented a complete rout.

Casualties were heavy: Union losses totaled over 14,000; Confederate losses were about 7,200. Despite higher proportional losses, the victory was decisive for Lee.

The battle humiliated Pope (who was soon relieved) and boosted Southern morale to new heights. It allowed Lee to shift the war's momentum, leading directly to the Maryland Campaign and the Battle of Antietam. For the Union, it exposed deep command problems and forced a reorganization under Maj. Gen. George B. McClellan.

Lee's Key Decisions and Brilliance

1. **Bold Division of His Army:** With only about 62,000 men against Pope's force which was roughly equal, Lee took an enormous risk by splitting his army twice: First, he sent Stonewall Jackson with ~24,000 men on a daring 56-mile flanking march around Pope's right flank (August 25–27). Jackson's "foot cavalry" captured the massive Union supply depot at Manassas Junction and then disappeared into the woods at Stony Ridge near the old First Bull Run battlefield.

   Lee then marched the remainder of his army (Longstreet's wing) north to reunite with Jackson — marching in the open while separated from half his force by the entire Union army.

2. **Deception and Speed:** Lee's coordination with Jackson (via cavalry commander J.E.B. Stuart) was flawless despite primitive communications. Pope was completely fooled into believing the Confederates were retreating, when in reality Jackson was already behind him.

3. **The Trap on August 29–30:** On August 29 Jackson's men repulsed repeated Union assaults along an unfinished railroad cut in a defensive masterpiece. The next day August 30 when Pope prematurely concluded Jackson was retreating and launched a poorly coordinated attack, Lee unleashed Longstreet's 28,000-man wing in the largest simultaneous assault of the war up to that point — a devastating 5-mile-wide counterattack that smashed into Pope's exposed left flank.

Results

Pope's army was shattered and driven back to Washington in a single campaign.

The victory emboldened Lee to launch his first invasion of the North (Maryland Campaign), leading directly to Antietam.

Why Lee's Performance Was Exceptional

1. He repeatedly divided his smaller army in the face of a larger enemy—a gamble that would have ended his career if Pope had shown any aggressiveness.
2. Perfect coordination between Jackson (independent action for days) and Longstreet.
3. Superb timing of Longstreet's massive counterstroke on August 30.
4. Complete moral and strategic dominance over John Pope, whom Lee outgeneraled at every turn.

Historians often rank Second Bull Run as Lee's most perfectly executed campaign. James M. McPherson calls it "Lee's most clear-cut victory," and many consider it the high-water mark of Lee's reputation as a daring offensive commander. It showcased his trademark traits: audacity, rapid maneuver, willingness to accept great risk for decisive results, and an uncanny ability to read his opponent's mind.

Casualties

Due to Lee's tactical brilliance, Union casualties of 14,000 were double that of the 7,000 the Confederates lost. Pope lost almost a third of his army.

Second Battle of Bull Run Casualty Comparison											
Union						Confederate					
Strength	Killed	Wounded	Captured/ Missing	Total	%	Strength	Killed	Wounded	Captured/ Missing	Total	%
50,000	1,742	8,452	4,263	**14,457**	**29%**	62,000	1,096	6,202	NA	**7,298**	**12%**

Rating Lee's Performance at the Second Battle of Bull Run

The Second Battle of Bull run was a Lee masterpiece. Strategically he accomplished his objective of protecting Confederate supply/rail lines and ridding Northern Virginia of the Union Army. Tactically Lee outgeneraled Pope, often dividing his forces to land blows from several directions. Lee inflicted double the casualties of his own. A resounding victory deserving a perfect *Excellent* score of 5.0.

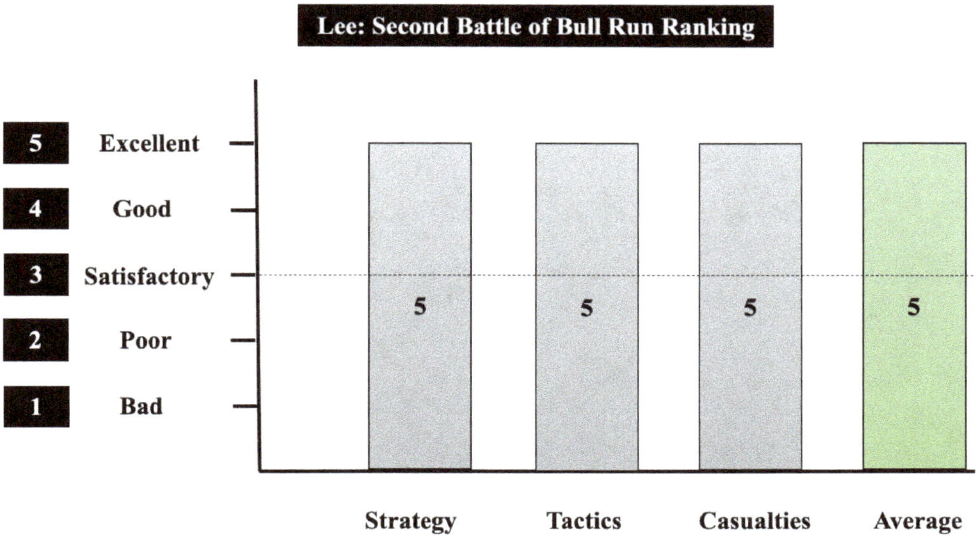

Second Battle of Bull Run

Map by Hal Jespersen, www.cwmaps.com

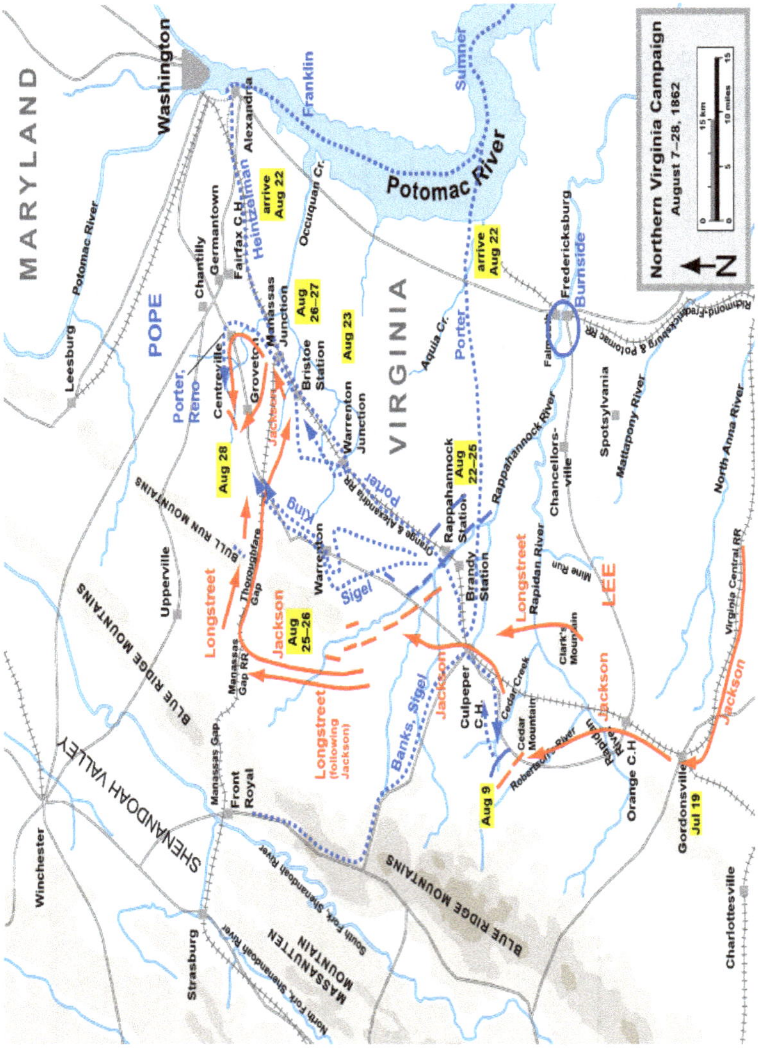

Lee: *Battle of South Mountain* (September, 1862)

Robert E. Lee's performance at South Mountain is generally regarded by historians as one of the weakest and most problematic episodes of his military career. It was part of the Maryland Campaign that immediately preceded Antietam, and it came at the worst possible moment for the Confederacy because of an infamous Lost Order (Special Orders No. 191) that was found and used against him by the Union army.

Synopsis

The **Battle of South Mountain**, fought on September 14, 1862, in Frederick and Washington Counties, Maryland, was the first major engagement of Lee's 1862 Maryland Campaign. It occurred three days after a Union soldier came upon a carelessly discarded Confederate Order of Battle (Special Orders 191) which revealed that the Confederate Army of Northern Virginia was dangerously divided into several parts.

Maj. Gen. George B. McClellan, commanding the Army of the Potomac (about 28,000 engaged), moved aggressively to exploit the opportunity and prevent Lee from reuniting. Lee had only a small force—roughly 18,000 men under Maj. Gen. D.H. Hill—to defend the critical passes through South Mountain: Fox's Gap, Turner's Gap (and nearby Frostown Gap), and Crampton's Gap farther south.

Fighting erupted in the morning and raged through the afternoon and evening across steep, rugged terrain. At **Fox's Gap**, Union IX Corps troops under Jacob D. Cox overran Brig. Gen. Samuel Garland's brigade, mortally wounding Garland, but reinforcements allowed the Confederates to hold the western edge as night fell. Fierce combat also occurred around Turner's Gap, where Union I Corps attacked and Maj. Gen. Jesse L. Reno (IX Corps commander) was killed. At **Crampton's Gap**, Union VI Corps under William B. Franklin launched a powerful assault in the late afternoon, overwhelming a small Confederate force under Howell Cobb and securing the pass.

By nightfall, Union forces had seized all three gaps, forcing the outnumbered Confederates to withdraw westward. McClellan reported a "glorious victory," and the battle provided a significant morale boost to the previously defeated Army of the Potomac.

Casualties were heavy for a single day's fighting: Union losses totaled 2,325; Confederate losses were higher at 2,685.

Although a clear Union tactical victory, the day-long delay allowed Lee precious time to concentrate his scattered divisions near Sharpsburg along Antietam Creek. McClellan's characteristic caution on September 15 prevented him from destroying Lee's army in detail. The battle set the stage for the far bloodier Battle of Antietam three days later and marked the initial check on Lee's first invasion of the North.

Lee's Decisions and Mistakes

1. **The Lost Order catastrophe** – While not directly Lee's "fault" that the order was lost, the very decision to split his army so widely in enemy territory with detailed written orders was extremely risky. Most commanders (including Lee himself later in the war) avoided such specificity in writing when operating on enemy soil.

2. **Initial decision to retreat, then reversal** – After learning McClellan had the order, Lee first ordered a retreat across the Potomac, then countermanded it and decided to fight at South Mountain. This flip-flop wasted crucial hours and left his defending forces with almost no time to prepare adequate defenses.

3. **Underestimating McClellan** – Lee assumed (as he often did) that McClellan would move slowly even with perfect intelligence. In fact, McClellan moved unusually aggressively on the 14th, forcing the crisis at the gaps.

4. **Inadequate forces committed** – Lee left only a fraction of his army to hold the mountain passes while the bulk (Jackson, McLaws, Walker) was tied down at Harpers Ferry. Even when it became clear the gaps were under heavy assault, Lee sent only limited reinforcements (mostly Hood's and some of Longstreet's troops arriving late).

5. **Almost lost the entire campaign** – By nightfall on September 14, the Army of Northern Virginia was in serious danger of being cut in half or destroyed in detail. Only McClellan's habitual caution in not pursuing aggressively after dark, combined with Jackson's rapid success at Harpers Ferry the next morning (Sept 15), allowed Lee to concentrate again at Sharpsburg.

Historians' Verdict

Stephen W. Sears (Landscape Turned Red and other works): Calls South Mountain "Lee's worst day of the war" and says the campaign came within hours of total collapse.

Joseph L. Harsh (Taken at the Flood): While more sympathetic to Lee, still describes the defense of the gaps as improvised, poorly coordinated, and nearly disastrous.

James M. McPherson, Gary W. Gallagher, and most modern scholars view South Mountain as the clearest example of Lee taking excessive risk that almost ended the Maryland Campaign on September 14 itself.

In Summary

Lee's performance at South Mountain was uncharacteristically poor: overconfident planning, a catastrophic intelligence failure, vacillation once the danger was known, and committing far too few troops to a vital delaying action. The fact that he ultimately salvaged the situation and fought at Antietam three days later speaks to his resilience and the extraordinary fighting quality of his army, but September 14, 1862, remains the closest Robert E. Lee ever came to outright strategic defeat during the war.

Casualties

Lee's casualties were higher than the those of the Union at 15% of total compared to 8%.

Battle of South Mountain Casualty Comparison											
Union						Confederate					
Strength	Killed	Wounded	Captured/ Missing	Total	%	Strength	Killed	Wounded	Captured/ Missing	Total	%
28,000	443	1,807	75	**2,325**	**8%**	18,000	325	1,560	800	**2,685**	**15%**

Rating Lee's Performance at the Battle of South Mountain

Lee's performance at the Battle of South Mountain overall was *Poor* with a score of 2.33. Strategically (Score: 2) he failed to combine his dispersed forces once it was known the Union had his plans in their pocket. Tactically (also 2) he attempted to defend the mountain gaps with too few troops compared to the known Union strength in the area. Lee's casualties were almost double the Union percentage wise which is unusual for an army on the defensive. Just a poorly thought out and fought battle by Lee. We all have bad days, and this was one of Lee's.

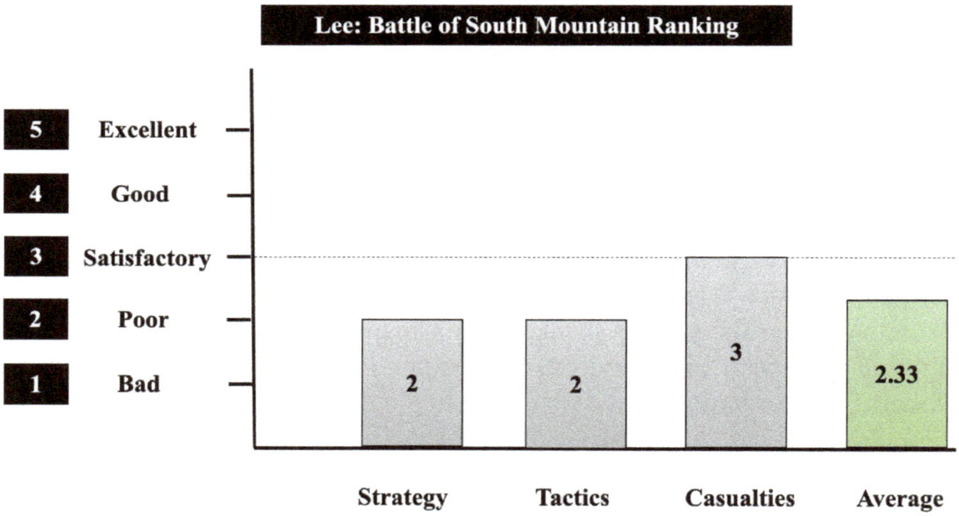

Battle of South Mountain

Map Courtesy of the Civil War Trust

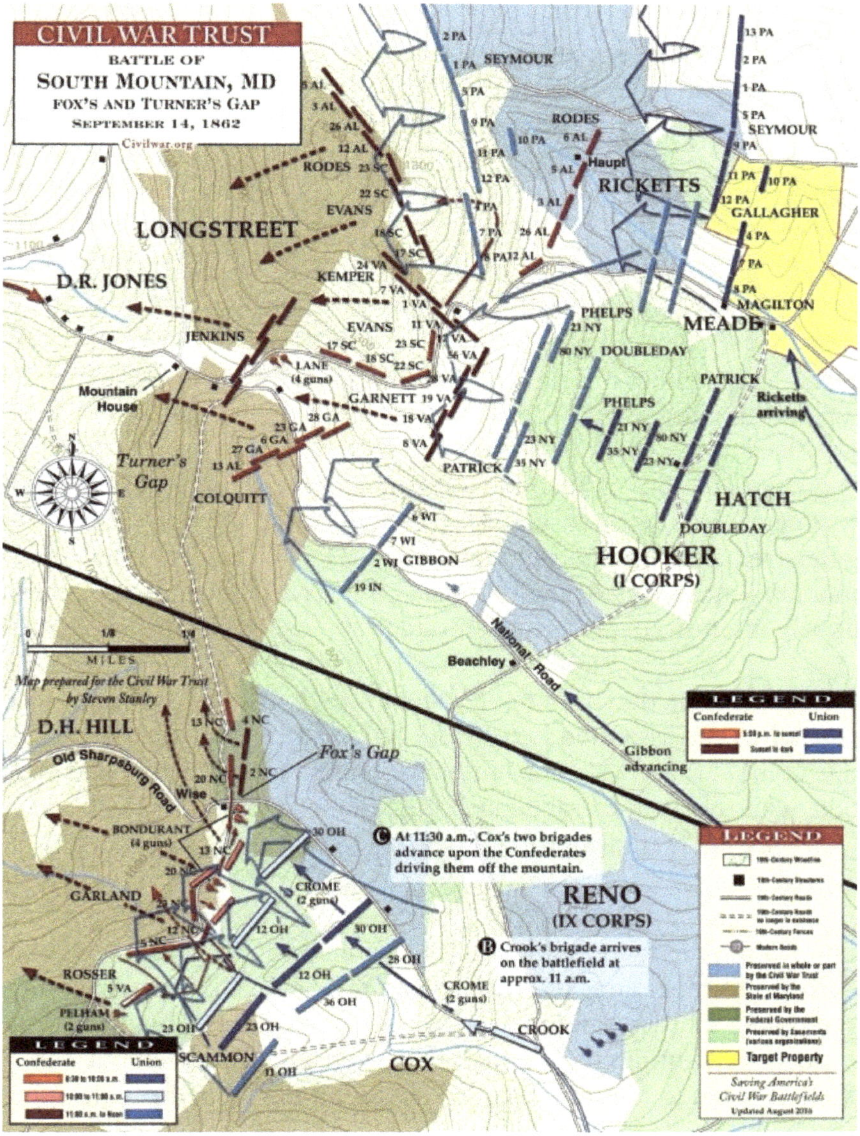

Lee: *Battle of Antietam* (September, 1862)

Robert E. Lee's performance at Antietam is one of Lee's most debated command performances of the Civil War. It marked the end of Lee's first invasion of the North (the Maryland Campaign).

Synopsis

The **Battle of Antietam** (also called the **Battle of Sharpsburg** in the South), fought on September 17, 1862, near Sharpsburg, Maryland, remains the bloodiest single day in American military history. It was the climactic engagement of Lee's first invasion of the North during the Maryland Campaign.

Following his victory at Second Bull Run and the capture of Harpers Ferry, Lee crossed into Maryland with roughly 30,000 men (many stragglers reduced effective strength). Maj. Gen. George B. McClellan, commanding the Army of the Potomac (about 53,000 engaged), pursued cautiously after discovering Lee's divided forces via Special Orders 191.

Fighting erupted at dawn. In the northern sector, Union Maj. Gen. Joseph Hooker's I Corps assaulted Lee's left flank near the Dunker Church. Savage combat raged through Miller's Cornfield and the East and West Woods, with repeated Confederate counterattacks by "Stonewall" Jackson's men. By mid-morning, over 8,000 casualties had fallen in this small area.

Around 9 a.m., the focus shifted to the center. Union forces under Maj. Gen. Edwin Sumner attacked Maj. Gen. D.H. Hill's division defending the sunken road—thereafter known as "Bloody Lane." After hours of brutal, close-quarters fighting, the Confederates were finally driven out, but McClellan failed to commit reserves to exploit the breakthrough.

In the afternoon, Union Maj. Gen. Ambrose Burnside's IX Corps finally crossed Antietam Creek on the southern flank and began rolling up Lee's right. Just as the Confederate line seemed about to collapse, Maj. Gen. A.P. Hill's division arrived after a rapid 17-mile march from Harpers Ferry and launched a fierce counterattack that saved the day.

By 5:30 p.m., the fighting ended with both armies still holding roughly their original positions. Total casualties exceeded 22,700: Union losses were over 12,000; Confederate losses totaled about 10,000.

Tactically, the battle was a bloody stalemate, but strategically it was a Union victory. Lee's invasion was halted, and he withdrew across the Potomac into Virginia on September 18–19. McClellan's failure to pursue aggressively drew sharp criticism.

The battle's greatest significance came five days later when President Abraham Lincoln, using the "victory" as political cover, issued the Preliminary Emancipation Proclamation. This transformed the war's purpose from preserving the Union to also ending slavery, discouraged European intervention on behalf of the Confederacy, and paved the way for the enlistment of Black soldiers.

Antietam marked a major turning point, boosting Northern morale while ending Lee's immediate offensive threat and setting the stage for the war's next phase.

Antietam: Union and Confederate Casualties at the Dunker Church

Strengths of Lee's Performance

1. **Extraordinary personal courage**: Lee was on the battlefield almost constantly, often under artillery fire, rallying troops.

2. **Masterful improvisation**: Despite having his plans handed to the enemy and his army divided, he concentrated in time to fight on ground of his own choosing.

3. **Defensive genius**: Positioned his army with Antietam Creek at his back (normally a terrible idea), but the terrain funneled Union attacks into killing zones (Cornfield, Sunken Road).

4. **Subordinate leadership**: Relied heavily on Stonewall Jackson, Longstreet, and later A.P. Hill — and they delivered.

Criticisms of Lee's Performance

1. **Overconfidence bordering on recklessness**: Fighting with his back to the Potomac River with only one viable crossing (Boteler's Ford) risked annihilation if he lost.

2. **Refusal to retreat even when the situation looked dire**: By late afternoon Lee had virtually no reserves left and was personally committing the last artillery batteries. Many historians believe a more prudent commander would have withdrawn across the Potomac the night of the 17th or morning of the 18th.

3. **The "lost order" disaster happened under his watch**: Dividing the army so aggressively while deep in enemy territory was an enormous gamble that almost paid off — but only because McClellan moved slowly.

4. **Casualties were catastrophic**: Roughly a third of Lee's army became casualties in one day (about 10,300 out of ~30,000 engaged). The Army of Northern Virginia never really recovered its pre-Antietam strength in veterans or morale.

The Day After (September 18)

McClellan, despite having fresh reserves (V and VI Corps barely touched), refused to renew the attack. Lee daringly stood his ground all day on the 18th, challenging McClellan to resume the fight. When McClellan declined, Lee withdrew across the Potomac that night in good order — a tactical draw that became a strategic Union victory because it halted the invasion and gave Lincoln the "victory" he needed to issue the Emancipation Proclamation.

Historical Verdict

Tactically: Lee achieved a defensive miracle with an outnumbered, exhausted army and probably fought McClellan to a draw on the day.

Strategically: The Maryland Campaign was a clear failure

Many modern historians (e.g., Stephen Sears, James M. McPherson) rate Antietam as Lee's worst-handled battle in terms of risk-reward calculation, while still acknowledging his personal bravery and the remarkable performance of his subordinates.

In short: Lee's performance at Antietam was audacious, almost reckless, tactically impressive on the day, but strategically costly — one of the clearest examples of his tendency to take enormous risks that sometimes paid huge dividends (Chancellorsville, Second Manassas) and sometimes came within a hair's breadth of disaster (Antietam).

Casualties

Casualties on both sides exceeded 22,000—the bloodiest one day of battle in the war. Lee's casualties were slightly higher percentage wise at 34% compared to 23% for McClellan. At this time in the war, a loss of 10,000 men from Lee's army was a harbinger of the future.

Battle of Antietam Casualty Comparison

Union						Confederate					
Strength	Killed	Wounded	Captured/ Missing	Total	%	Strength	Killed	Wounded	Captured/ Missing	Total	%
53,632	2,108	9,549	753	**12,410**	**23%**	30,646	1,567	7,752	1,018	**10,337**	**34%**

Rating Lee's Performance at Antietam

Lee's performance at Antietam is mixed. Strategically his goal was to win a major victory in a northern state to 1) encourage northerners to turn against the war, and 2) provide fodder for England and France to recognize the Confederacy as a nation. Instead Lee was forced to flee back to Virginia with a badly decimated army. Thus, Lee failed in his strategic objective leading to a Strategy rating of Poor or 2.0. Tactically Lee's performance was Good, as he successfully fought to an essential draw a much larger

Union army. Lee's casualties while lower overall compared to the Union were still over one-third of his army—which just was not sustainable for the resource poor Confederacy. Overall score: rating of *Poor/Satisfactory* at 2.67.

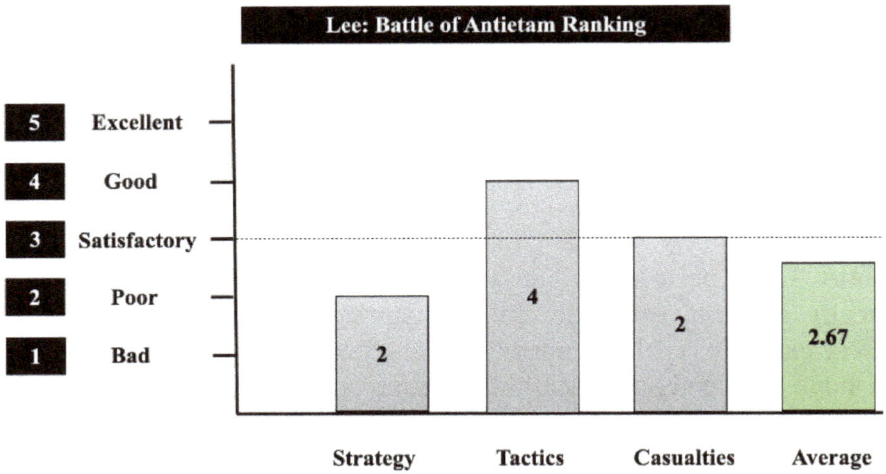

Battle of Antietam

Map by Hal Jespersen, www.cwmaps.com

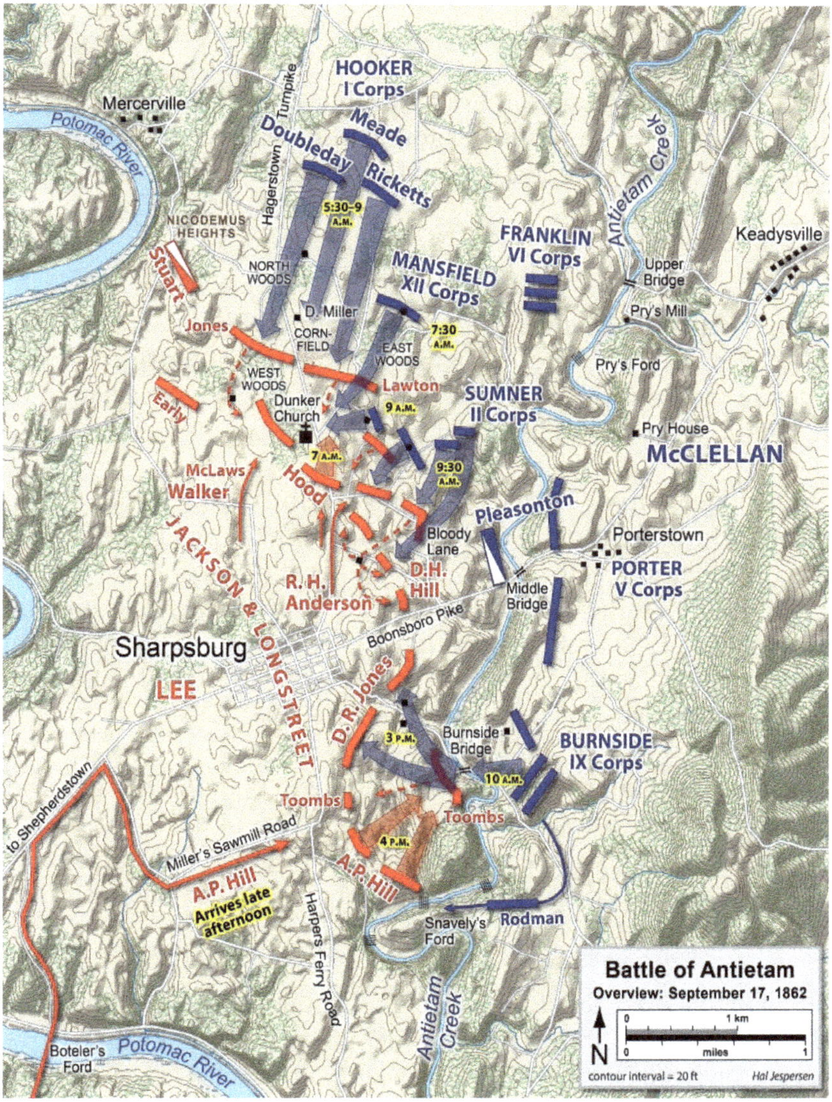

Lee: *Battle of Fredericksburg* (December, 1862)

Robert E. Lee's performance at Fredericksburg is widely regarded as one of his finest victories and a perfect example of defensive tactical warfare that might have served Lee well at later battles such as Gettysburg.

Synopsis

The **Battle of Fredericksburg**, fought December 11–15, 1862, in and around Fredericksburg, Virginia, was one of the most lopsided Confederate victories of the American Civil War and a devastating defeat for the Union Army of the Potomac.

After replacing George B. McClellan in November 1862, Maj. Gen. Ambrose E. Burnside (commanding ~114,000 men) planned a rapid advance on Richmond. He concentrated his army opposite Fredericksburg on the north bank of the Rappahannock River, intending to cross quickly and strike before Lee could concentrate his Army of Northern Virginia (~72,000 men). Delays in the arrival of pontoon bridges allowed Lee to fortify strong defensive positions south of the river.

On December 11, Union engineers began laying pontoon bridges under heavy fire from Confederate sharpshooters hidden in the town. After a massive artillery bombardment of Fredericksburg, Union troops crossed and occupied the ruined town on December 11–12. The main assault occurred on December 13.

Burnside launched a two-pronged attack. On the Union left, Maj. Gen. William B. Franklin's forces made some progress against Lt. Gen. Thomas "Stonewall" Jackson's corps near Hamilton's Crossing but were ultimately repulsed. On the right, repeated frontal assaults against Lt. Gen. James Longstreet's corps on Marye's Heights proved catastrophic. Union divisions advanced across an open 600-yard plain dominated by a sunken road fronted by a stone wall, backed by artillery on the heights. Wave after wave of Federals (fourteen charges in all) were cut down by devastating rifle and cannon fire. The "Slaughter Pen" in front of the stone wall became a scene of horrific carnage. One Confederate observer noted it was "not war but murder."

By nightfall on December 13, the Union attacks had failed completely with no ground gained. Burnside, deeply shaken, considered renewing the assault the next day but was dissuaded by his corps commanders. On the night of December 15, the Army of the Potomac withdrew across the river to Falmouth under cover of darkness and a storm.

Casualties were heavily one-sided: Union losses totaled 12,653; Confederate losses were 5,377.

The defeat was a severe blow to Northern morale and led to Burnside's replacement by Joseph Hooker in January 1863. Lee's defensive masterpiece boosted Confederate confidence, though the victory did not lead to further strategic gains. The battle highlighted the growing power of defensive positions against frontal infantry assaults in the age of rifled muskets.

General Ambrose Burnside

Union Attack at Fredericksburg

Why Lee's generalship was brilliant at Fredericksburg

1. **Superb use of terrain:** Lee positioned his army along a naturally strong defensive line: Marye's Heights (Longstreet's First Corps): A sunken road and stone wall at the base of the hill gave Confederate infantry perfect cover and enfilading fields of fire. Prospect Hill (Jackson's Second Corps) on the southern end: Anchored on the RF&P Railroad and marshy ground that channeled Union attacks.

2. **Perfect coordination between wings:** Despite Jackson's sector almost breaking on December 13 when Meade's division temporarily pierced the line, Lee remained calm, quickly reinforced with Early's division, and restored the position. Longstreet's front at Marye's Heights proved virtually impregnable (14 successive Union brigades were shredded in frontal assaults).

3. **Psychological dominance:** Lee famously remarked while watching the slaughter at the stone wall: "It is well that war is so terrible, otherwise we should grow too fond of it." This reflects both his tactical success and the grim realization of the carnage his dispositions caused.

4. **Strategic restraint**: After routing Burnside, Lee chose not to pursue aggressively across the Rappahannock River in winter. While some modern historians criticize this as a missed opportunity, most contemporaries and later analysts agree that a river crossing against

a still-large Union army in December carried enormous risks for little strategic gain.

Criticisms (minor in context)

Some Monday-morning quarterbacks argue Lee should have counterattacked more aggressively after repulsing the Union assaults. Jackson wanted to launch a night attack on December 13, which Lee vetoed (probably wisely, given the confusion night attacks often caused in that era).

Historical consensus

Fredericksburg is usually cited as Lee's most complete tactical masterpiece. Burnside himself admitted he played directly into Lee's hands by launching frontal assaults against prepared defenses. Military historians frequently rank it alongside the Roman's Cannae or Napoleon's Austerlitz as an example of near-perfect defensive warfare.

In short: At Fredericksburg, Lee fought the battle exactly as he wanted, on ground of his choosing, and inflicted one of the most lopsided defeats of the war. It remains the high-water mark of his reputation as a defensive tactician.

Casualties

Total Union casualties of over 12,000 were more than double Confederate losses of 5,377. This reflected Lee's successful defensive tactics behind fixed fortifications.

Battle of Fredericksburg Casualty Comparison

Union						Confederate					
Strength	Killed	Wounded	Captured/ Missing	Total	%	Strength	Killed	Wounded	Captured/ Missing	Total	%
114,000	1,284	9,600	1,769	**12,653**	**11%**	72,500	608	4,116	653	**5,377**	**7%**

Rating Lee's Performance at Fredericksburg

Lee's performance at Fredericksburg rates *Excellent* in all measures for a perfect 5.0 overall score. Strategically Lee accomplished his objective of blunting yet another Union incursion into Virginia roundly defeating

Burnside. Tactically his use of terrain and the flawless coordination of the wings of his army drove his victory.

Confederate casualties were less than half that of the Union. This was the type of defensive warfare that if maintained and continued might have eventually resulted in an overall stalemate in the Civil War, leading Americans and Europeans to tire of the war thus accomplishing the Confederate objective of independence.

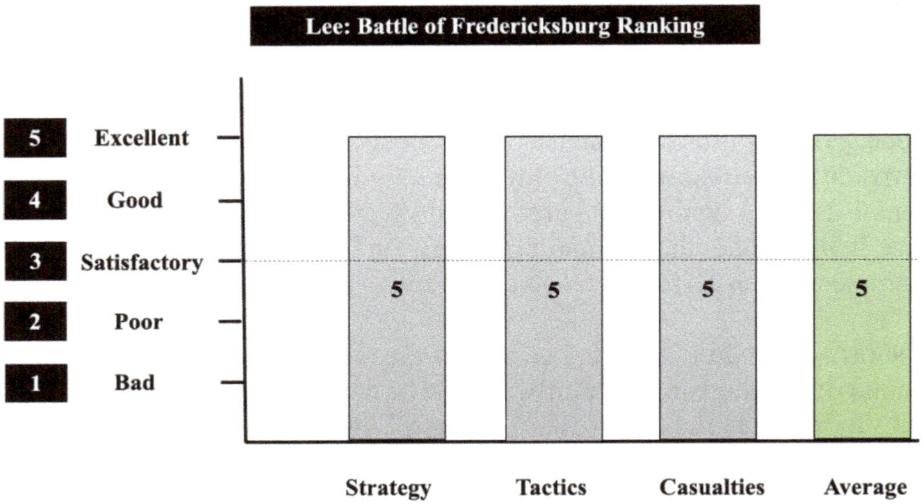

David W. Manly

Battle of Fredericksburg

Map by Hal Jespersen, www.cwmaps.com

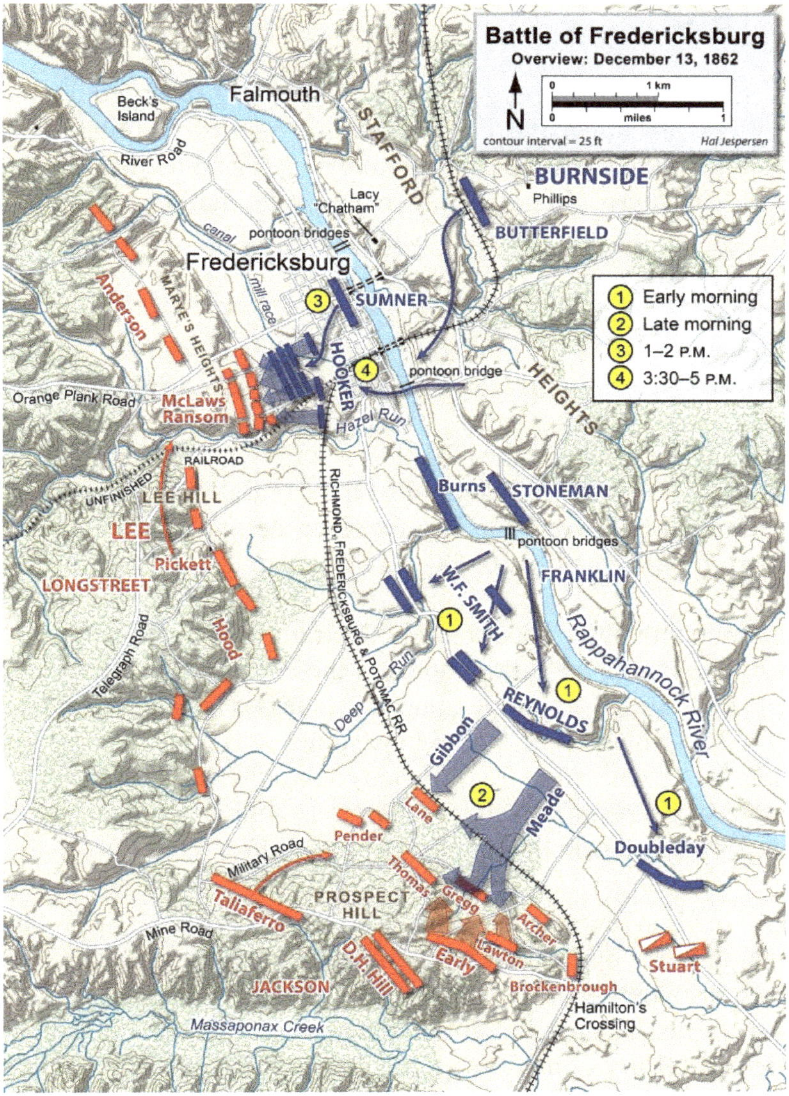

Grant: *Vicksburg Campaign* (October, 1862-July, 1863)

Ulysses S. Grant's Vicksburg Campaign is widely regarded as one of the most brilliant and decisive campaigns in American military history. It effectively split the Confederacy in two by capturing the Mississippi River fortress of Vicksburg and gave Grant even greater national prominence that eventually led to his appointment as general-in-chief.

Synopsis

The **Vicksburg Campaign** (December 1862–July 4, 1863) was a masterful Union operation led by Grant that secured control of the Mississippi River and split the Confederacy in two. Vicksburg, Mississippi, known as the "Gibraltar of the Confederacy," was heavily fortified on high bluffs overlooking the river, defended by Lt. Gen. John C. Pemberton with roughly 30,000–33,000 men.

Early attempts in winter 1862–63 failed. Grant tried canal digging, bayou expeditions, and a direct assault north of the city at Chickasaw Bayou, all thwarted by geography, weather, and Confederate resistance. In late March 1863, Grant adopted a bold new strategy. He marched his Army of the Tennessee (eventually ~75,000 men) down the west bank of the Mississippi while Admiral David D. Porter's gunboats ran past Vicksburg's batteries on the night of April 16–17.

On April 30, Grant crossed the river at Bruinsburg, Mississippi—the largest amphibious operation in American history up to that time. Cutting loose from his supply base, he launched a lightning campaign of maneuver. In just 17 days, his forces marched 180 miles and won five quick victories: **Port Gibson** (May 1), **Raymond** (May 12), **Jackson** (May 14, where they defeated Joseph E. Johnston's relief force and burned the state capital), **Champion Hill** (May 16, the campaign's decisive battle), and **Big Black River Bridge** (May 17). These actions prevented Pemberton and Johnston from uniting and drove Pemberton's battered army back into Vicksburg's defenses.

Grant arrived before the city on May 18. Two bloody assaults on May 19 and 22 failed with heavy Union losses, convincing him to lay siege. For 47 days, Union troops dug trenches, sapped forward, and subjected the city to constant artillery and naval bombardment. Civilians and soldiers endured starvation, living in caves as food supplies ran out and disease spread.

On July 4, 1863—one day after the Confederate defeat at Gettysburg—Pemberton surrendered the city and his entire garrison of 29,495 men. Union casualties for the full campaign totaled about 7,800; Confederate losses were far higher when including the surrender: over 36,000 total.

The victory gave the Union complete control of the Mississippi River, severed Texas, Louisiana, and Arkansas from the eastern Confederacy, and opened the way for further advances. President Lincoln declared the river "again goes unvexed to the sea." Grant's success, achieved through daring maneuver, persistence, and logistics, established him as the Union's premier commander and marked a major turning point in the war.

Union Gunboats Running the Blockade at Vicksburg

Key Phases and Grant's Decisions

Early Failures (Winter 1862–1863)

Grant attempted several conventional approaches that all failed:

- An overland advance through northern Mississippi that was blocked by a raid on his supply base at Holly Springs by Confederate General Van Dorn.
- Canal projects across the Mississippi peninsula opposite Vicksburg which would allow Union gunboats to bypass the heavily fortified city (abandoned due to flooding and Confederate fire).
- Attempts to attack Vicksburg from the north via a series of lakes/rivers/swamps (Yazoo Pass, Steele's Bayou, and Lake Providence expeditions) all of which bogged down.

Rather than becoming discouraged, Grant used these efforts to mislead the Confederates about his intentions and to keep his army active and trained.

The Bold Strategic Decision (April 1863)

Grant made the audacious choice to march his army down the west bank of the Mississippi, run past the Vicksburg batteries with Admiral Porter's gunboats and transports at night (April 16 & 22), then cross the river south of Vicksburg with no supply line back to Union territory.

This was an enormous risk: if defeated, his army could have been trapped against the river with no retreat. Lincoln and Halleck were extremely nervous, but Grant had complete confidence in the plan. Even though the Confederates in Vicksburg saw the Union boats passing by at night their gunfire was ineffective at halting the passage.

Operational Maneuver (April 30 – May 14, 1863)

Now south of Vicksburg on the western side of the Mississippi and after crossing at Bruinsburg (April 30), Grant cut loose from his supply base and lived off the country — a classic raid-turned-campaign. Five victories in 18 days:

- Port Gibson (May 1)
- Raymond (May 12)
- Jackson (May 14) — captured the state capital and destroyed Confederate rail and supply infrastructure
- Champion Hill (May 16) — decisive tactical victory that broke Pemberton's army
- Big Black River Bridge (May 17)

These battles drove Pemberton back into the Vicksburg defenses and isolated him from Joseph E. Johnston's relief force.

Siege of Vicksburg (May 18 – July 4, 1863)

After two costly assaults (May 19 and May 22) convinced Grant that the fortifications were too strong, he calmly shifted to a formal siege.

Masterfully coordinated army-navy-artillery-engineer operations, constant mining and sapping, and tightened the noose until the garrison (starving and demoralized) surrendered on July 4, 1863.

Captured ~29,500 Confederate soldiers, 172 cannon, and ~60,000 small arms with minimal additional Union losses.

Why Grant's Performance Was Exceptional

How Grant's Performance at Vicksburg was Exceptional		
Aspect	**Grant's Approach**	**Result/Impact**
Strategic Vision	Recognized that seizing Vicksburg required bold risk south of the city	Completely outmaneuvered Confederate Command
Flexibility	Abandoned failed plans quickly, adapted to circumstances	Turned winter "failures" into strategic deception
Logistics	Army lived of the land for 20 days—a rarity in warfare at the time	Kept army mobile enabling deception and surprise of the enemy
Leadership & Morale	Maintained the Army's confidence despite earlier set-backs	His soldiers trusted him, morale remained high
Coordination	Seamless coordination with Sherman and Admiral Porter of the Navy	All plans came together for "knock-out" blows. The Navy's cooperation was decisive

Contemporary and Historical Judgment

Lincoln after Vicksburg: "Grant is my man, and I am his the rest of the war."

British military historian J.F.C. Fuller called the campaign "one of the most perfectly executed campaigns in history."

Modern historians (e.g., Michael Ballard, Edwin Bears, Steven Woodworth) consistently rank it among the top two or three campaigns of the entire war (alongside Atlanta and perhaps Second Manassas).

In short, Grant's Vicksburg Campaign showcased every strength that would later make him the preeminent Union general: strategic insight, willingness to take calculated risks, operational speed, tactical tenacity, and an unbreakable will. It was the making of Grant's reputation and a mortal blow to the Confederacy.

Casualties

Grant's army in the Vicksburg campaign essentially "rolled up" the Confederate forces that opposed them. Confederate losses in the battles leading up to the siege of Vicksburg were two to three times the losses Grant suffered. And, of course, the entire garrison at Vicksburg was captured resulting in over 36,000 total losses compared to only 7,854 for the Union.

Vicksburg Campaign Casualty Comparison

	Union						Confederate					
	Strength	Killed	Wounded	Captured/ Missing	Total	%	Strength	Killed	Wounded	Captured/ Missing	Total	%
Raymond	11,000	NA	NA	NA	446	4%	3,500	NA	NA	NA	515	15%
Jackson	11,500	41	251	7	299	3%	6,000	NA	NA	NA	250	4%
Champion's Hill	32,000	410	1,844	187	2,441	8%	22,000	331	1,018	2,441	3,790	17%
Big Black River	12,370	39	297	3	279	2%	3,000	NA	NA	NA	1,751	58%
Siege of Vicksburg	77,000	766	3,793	276	4,835	6%	33,000	1,000	2,202	29,495	32,697	99%
Total		1,256	6,125	473	7,854	5%		1,331	3,220	31,936	36,487	100%

Rating Grant's Performance in the Vicksburg Campaign

Grant's performance in the Vicksburg Campaign rates a perfect 5.0 *Excellent* score. Grant's strategic objective was to open up the entire Mississippi Valley to the Union and shut down the use of the river to the Confederacy. Capturing Vicksburg accomplished this objective. Grant's tactics of cutting loose of his supply base and defeating the Confederate armies piecemeal was brilliant. Grant's army captured an entire Confederate army of over 30,000 men at a loss of under 8,000 on the Union side. It is no wonder that Grant's Vicksburg Campaign is still studied at war colleges worldwide.

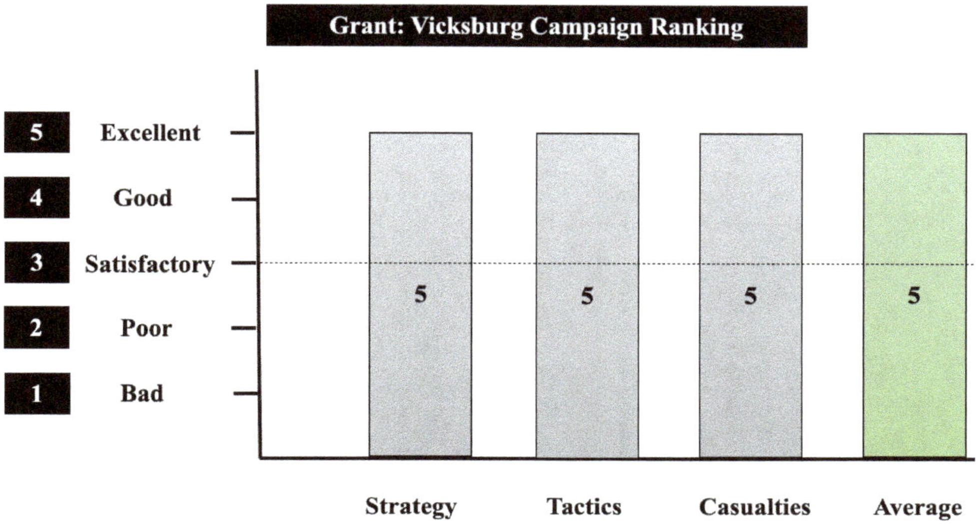

David W. Manly

The Vicksburg Campaign of 1863

Map by Hal Jespersen, www.cwmaps.com

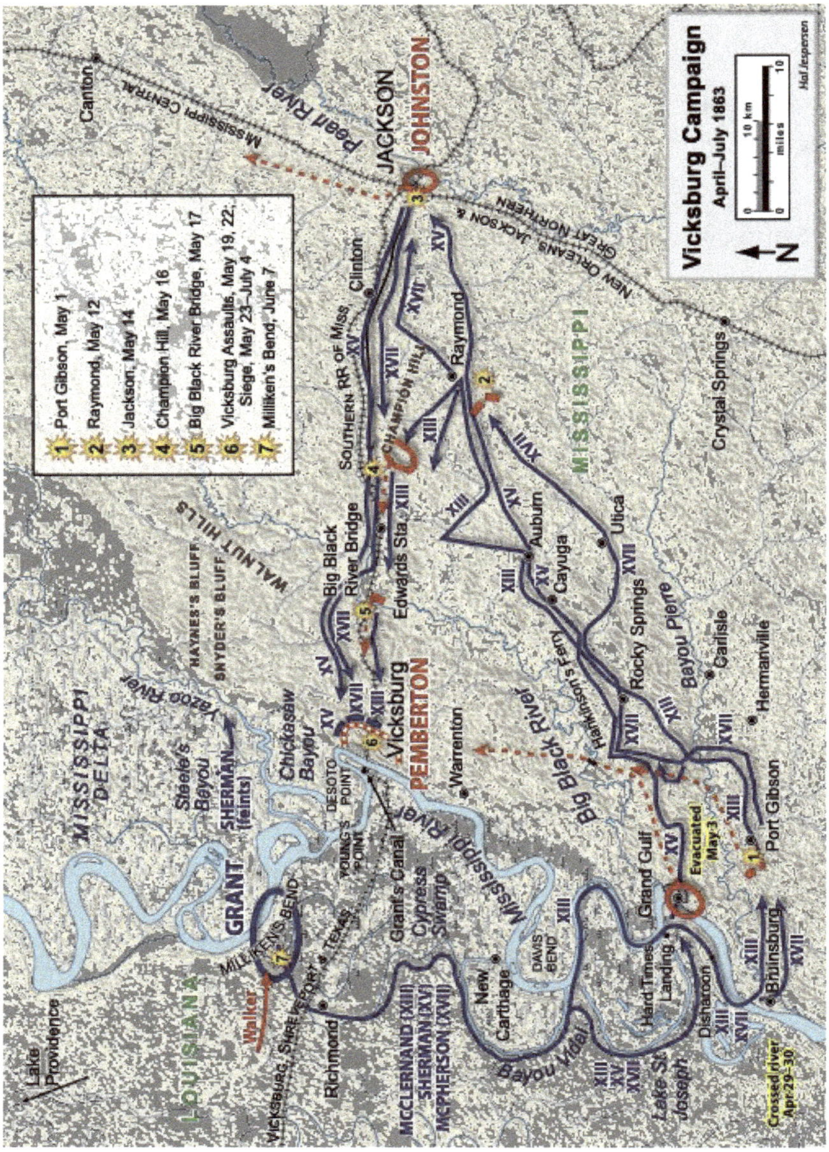

Lee: *Battle of Chancellorsville* (April, 1863)

Robert E. Lee's performance at the Battle of Chancellorsville (April 30 – May 6, 1863) is almost universally regarded as the masterpiece of his military career and one of the most brilliant performances by any general in history.

Synopsis

The **Battle of Chancellorsville**, fought April 30–May 6, 1863, in Spotsylvania County, Virginia, is widely regarded as Robert E. Lee's greatest tactical victory of the Civil War, often called his "perfect battle." Facing Maj. Gen. Joseph Hooker's much larger Army of the Potomac (approximately 133,000 men), Lee commanded only about 60,000 troops of the Army of Northern Virginia.

After the Union disaster at Fredericksburg, Hooker replaced Ambrose Burnside and devised an ambitious plan to outflank Lee. He left a strong force under Maj. Gen. John Sedgwick at Fredericksburg as a diversion while crossing the Rappahannock River upstream with the main army, concentrating near the Chancellorsville crossroads in the dense Wilderness. By May 1, Hooker had achieved a strong position and held numerical superiority.

Lee, defying military doctrine, boldly divided his smaller army. He left a small force to confront Sedgwick and took the rest west to meet Hooker. On the evening of May 2, Lee further split his forces, sending Lt. Gen. Thomas "Stonewall" Jackson with about 28,000 men on a daring 12-mile flanking march around the Union right. Jackson's surprise attack at dusk routed the unsuspecting XI Corps, rolling up the Union flank in chaos and inflicting heavy casualties.

Tragically, while returning from reconnaissance that night, Jackson was accidentally shot by his own men (friendly fire). He suffered a severe wound to his left arm, which was amputated, but he later developed pneumonia and died on May 10.

Fighting continued fiercely on May 3 as Confederate forces under J.E.B. Stuart (temporarily replacing Jackson) pressed the attack. Hooker, stunned by the flank attack and suffering a minor injury when a cannonball struck a pillar near him, grew increasingly cautious and defensive. Meanwhile, Sedgwick's force crossed at Fredericksburg and fought at Second Fredericksburg and Salem Church but failed to break through to link with Hooker.

By May 5–6, with his army bloodied and morale shaken, Hooker ordered a full retreat across the Rappahannock River, abandoning the campaign.

Casualties were severe: Union losses totaled about 18,000; Confederate losses were approximately 13,000.

Though a brilliant victory that humiliated Hooker and boosted Southern morale, Chancellorsville came at a terrible cost. The irreplaceable loss of Stonewall Jackson deprived Lee of his most trusted and aggressive lieutenant. The battle set the stage for Lee's second invasion of the North, leading directly to Gettysburg two months later.

General Joseph Hooker

Why it is considered an extraordinary performance by Lee:

1. **Extreme numerical disadvantage**: Lee faced Joseph Hooker's Army of the Potomac with roughly 60,000 Confederates against

roughly 133,000 Federals — worse than 2-to-1 odds overall, and in some phases closer to 4- or 5-to-1 on individual sectors of the field.

2. **Audacious division of his already smaller army — twice**: When Hooker advanced, Lee left only 10–12,000 men under Jubal Early to hold the Fredericksburg heights against John Sedgwick's 40,000+ while marching west with the bulk of his army to meet Hooker at Chancellorsville. Then, after learning Hooker's main force was in the Wilderness around Chancellorsville, Lee committed what is still taught in military academies as one of the boldest moves of the war: on May 1–2 he detached Stonewall Jackson with 28,000 men (nearly half of Lee's remaining strength) on a 12-mile flanking march to attack Hooker's exposed right flank (XI Corps) while Lee personally held the front with only ≈ 17,000 men against 70,000+ Federals.

3. **Perfect coordination and timing**: Jackson's flank attack on May 2 at ~5:15 p.m. completely shattered Howard's XI Corps and rolled up the Union right in one of the most successful surprise attacks in American military history. The speed, secrecy, and execution of the march (screened by Stuart's cavalry) were flawless.

4. **Psychological dominance**: Lee completely out-generaled Hooker. Hooker had started the campaign with an excellent plan and high confidence ("May God have mercy on General Lee, for I will have none"), but Lee's aggressive maneuvers caused Hooker to lose his nerve, go on the defensive, and contract into the Wilderness where his artillery and numbers were neutralized.

5. **Tactical victory despite the odds**: In four days of fighting Lee inflicted over 18,000 Union casualties while suffering less than 13,000 of his own — a shocking casualty-exchange ratio given he was heavily outnumbered. He forced the Army of the Potomac to retreat back across the Rappahannock.

The famous quote that sums it up

After Jackson's attack succeeded on May 2, Lee reportedly said when he met Jackson that night: "Their right is gone — we have them!" And later, when told of Union confusion: "The enemy is in my power."

The one great blemish

The victory cost the irreplaceable Stonewall Jackson, mortally wounded by friendly fire on the night of May 2. Lee's heartfelt note to Jackson ("Could I have directed events; I should have chosen for the good of the country to be disabled in your stead") and his later remark after hearing of Jackson's death ("I have lost my right arm") show how dearly the victory was purchased.

Lee and Stonewall Jackson

Historical Verdict

West Point, the U.S. Army Command & General Staff College, and virtually every serious military historian rate Chancellorsville as Lee's tactical masterpiece.

British field marshal Lord Wolseley called it "the finest battle of modern times."

Even Union officers admitted it was a humiliating defeat; George G. Meade reportedly said, "If we can't beat Lee with these odds, we'd better go home."

In short, Chancellorsville is the battle that cemented Lee's reputation as one of history's greatest battlefield commanders — an almost miraculous victory won by audacity, perfect intelligence (Stuart's cavalry reconnaissance), trust in his subordinates, and an uncanny ability to read the enemy commander's mind.

Casualties

Total Union casualties of over 18,000 were almost 40% higher than Confederate losses. However, Lee's total casualties were 21% of his total forces, significantly higher than the Union percentage.

Battle of Chancellorsville Casualty Comparison

Union						Confederate					
Strength	Killed	Wounded	Captured/ Missing	Total	%	Strength	Killed	Wounded	Captured/ Missing	Total	%
133,868	1,606	9,762	6,919	**18,287**	**14%**	60,298	1,665	9,081	2,018	**12,764**	**21%**

Rating Lee's Performance at Chancellorsville

Lee deserves a perfect score of 5.0 for accomplishing his strategic objective of blunting Hooker's thrust to destroy his army and capture Richmond. Instead, the Union army was whipped and forced to march back north. Tactically Lee was also brilliant especially by directing Stonewall Jackson to detach and flank attack the Union army almost rolling up Hooker's entire command. The only very minor blemish on the battle is Lee's high percentage of losses which the Confederacy just could not maintain over time. Overall score: *Good to Excellent* or 4.67.

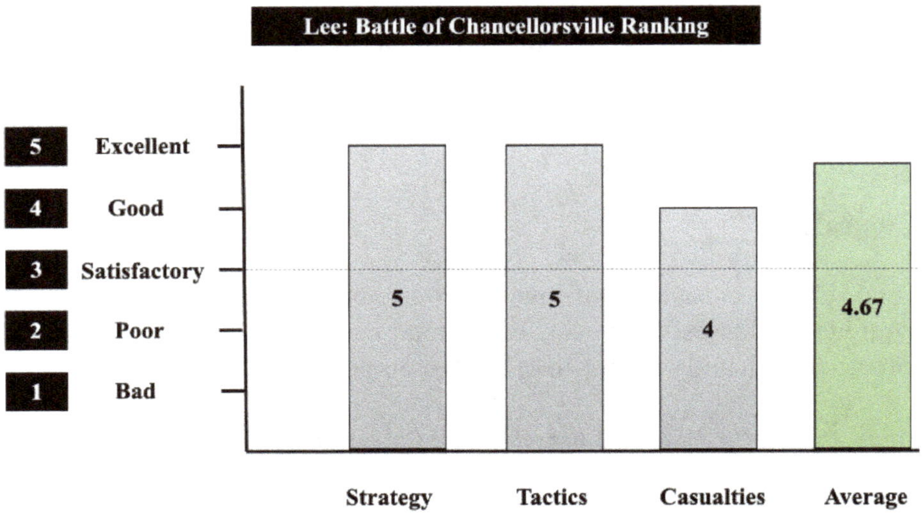
Lee: Battle of Chancellorsville Ranking
5 Excellent
4 Good
3 Satisfactory
2 Poor
1 Bad
5
5
4
4.67
Strategy
Tactics
Casualties
Average

David W. Manly

Battle of Chancellorsville

Map by Hal Jespersen, www.cwmaps.com

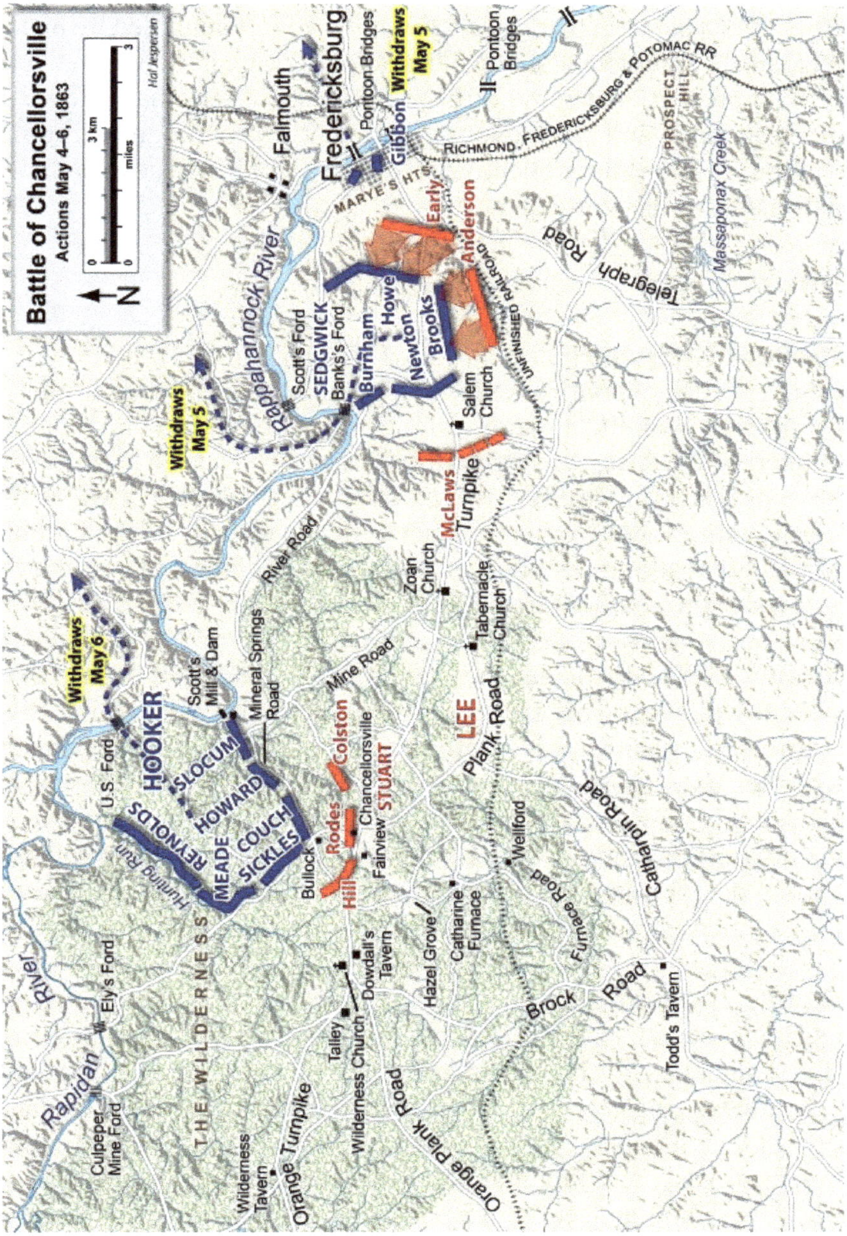

Lee: *Battle of Gettysburg* (July, 1863)

Robert E. Lee's performance at Gettysburg is widely regarded by historians as the clearest strategic and tactical defeat of his career. While Lee was not utterly incompetent, and some of his decisions had reasonable logic at the time, the campaign and battle exposed serious flaws in judgment, command style, and execution that collectively turned a bold invasion of the North into a disaster for the Confederacy.

Synopsis

The **Battle of Gettysburg**, fought July 1–3, 1863, in and around Gettysburg, Pennsylvania, was the largest and bloodiest battle of the American Civil War and a major turning point. Lee, commanding about 75,000 men of the Army of Northern Virginia, invaded the North hoping to relieve pressure on Virginia, win a decisive victory, and possibly force peace negotiations. Maj. Gen. George G. Meade's Army of the Potomac, roughly 104,000 strong, met him in an unplanned clash that became the war's decisive engagement in the Eastern Theater.

Day 1 (July 1): The battle began when Confederate forces under A.P. Hill advanced toward Gettysburg for supplies and collided with Union cavalry under Brig. Gen. John Buford. Buford's dismounted troopers delayed the Confederates until Union infantry arrived. Maj. Gen. John F. Reynolds was killed early. Fierce fighting west and north of town overwhelmed two Union corps. By evening, Union troops retreated through Gettysburg and took strong defensive positions on Cemetery Hill and Culp's Hill south of town. Lee failed to press the advantage that night.

Day 2 (July 2): Meade formed a strong "fishhook" defensive line from Culp's Hill, around Cemetery Hill, along Cemetery Ridge to Little Round Top. Lee launched coordinated assaults on both Union flanks. On the left, Lt. Gen. James Longstreet's corps attacked through the Peach Orchard, Wheatfield, Devil's Den, and toward Little Round Top. Intense fighting saw heroic stands, including the 20th Maine under Col. Joshua Chamberlain, who repulsed the 15th Alabama with a bayonet charge after running low on ammunition. On the right, Lt. Gen. Richard Ewell's attacks on Culp's Hill and East Cemetery Hill gained some ground but failed to dislodge the defenders. Heavy casualties mounted on both sides, but the Union line held.

Day 3 (July 3): After renewed fighting on Culp's Hill in the morning, Lee ordered a massive frontal assault on the Union center at Cemetery Ridge. Following a massive two-hour artillery barrage, about 12,500 Confederates—primarily Pickett's Division supported by Pettigrew's and Trimble's brigades—advanced across nearly a mile of open fields in what became known as Pickett's Charge. Exposed to devastating artillery and rifle fire, the attackers suffered terrible losses. A few hundred briefly breached the line near the "Angle" in hand-to-hand fighting (the "high-water mark of the Confederacy"), but Union reinforcements quickly sealed the gap. The charge collapsed with roughly 50% casualties among the attackers. Confederate cavalry attempts on the Union rear also failed.

On July 4, Lee began his retreat to Virginia under heavy rain. Meade did not pursue aggressively.

Casualties were staggering: Union losses totaled approximately 32,000; Confederate losses were also about 32,000. Total casualties reached almost 65,000, the highest of any single battle in American history.

The Union victory halted Lee's invasion, boosted Northern morale, and prevented potential foreign recognition of the Confederacy. Combined with the fall of Vicksburg the same day, it marked a strategic turning point. Four months later, President Lincoln delivered the Gettysburg Address at the battlefield cemetery dedication.

Union General George Meade

Key Criticisms of Lee's Performance at Gettysburg:

1. **Strategic Decision to Invade Pennsylvania**: After Chancellorsville (May 1863), Lee persuaded Jefferson Davis to approve a second invasion of the North. Intended goals: relieve Virginia farmland, gather supplies, threaten Washington/Baltimore, possibly detach Maryland, and force a decisive battle on Union soil. Risks were enormous: operating far from home with extended supply lines, no realistic hope of significant foreign recognition after Vicksburg's fall became likely, and facing a freshly reorganized Army of the Potomac under a competent new commander (George G. Meade). Most modern historians (e.g., Gary Gallagher, Stephen Sears, Edwin Coddington) believe the invasion was an unacceptable gamble with the Confederacy's best field army.

2. **Loss of Intelligence and Cavalry Screen**: Lee allowed J.E.B. Stuart to take the bulk of the Confederate cavalry on a controversial ride around the Union army (June 25–July 2. This left Lee virtually blind for over a week during the approach march. He had almost no idea where the Army of the Potomac was until June 28–29, far too late. Lee himself later wrote that Stuart's absence "made me very anxious" and was a major factor in the campaign's failure.

3. **Poor Coordination and Subordinate Selection on July 1–3:** Lee arrived on the field July 1 after the battle had already begun and issued vague, discretionary orders ("If practicable, attack") that led to a larger engagement than he originally did not intend. Corps commanders Richard Ewell and A.P. Hill performed poorly on July 1 (especially Ewell's failure to take Cemetery Hill when it was weakly held). James Longstreet, normally Lee's most reliable lieutenant, was sullen and slow on July 2 and 3 because he fundamentally disagreed with Lee's offensive plan. Lee failed to energize or overrule him decisively.

4. **The Decision to Attack Center on July 3 (Pickett's Charge)**: The most criticized single decision of Lee's career. After failing to turn either Union flank on July 2, Lee concluded that Meade had weakened his center to reinforce the flanks and ordered a frontal assault with ~13,000 men across 1,000+ yards of open ground against prepared positions on Cemetery Ridge. The attack was shredded by artillery and rifle fire; casualties approached 50–60%.

Virtually every modern historian (Sears, Pfanz, Trudeau, Hess, etc.) considers the attack hopeless from the moment it was conceived. Longstreet told Lee on the morning of July 3, "No 15,000 men ever arrayed for battle can take that position."

5. **Overconfidence and "Lee as Offense-Only General"**: Lee repeatedly told subordinates, "The enemy is there, and I am going to attack him there." This reflected an almost fatalistic belief that his army could achieve anything on the offensive—a mindset forged by earlier victories (Second Manassas, Chancellorsville) but catastrophically misplaced against Meade's strong defensive position.

Pickett's Charge

Mitigating Factors Sometimes Offered in Lee's Defense

Lee was suffering from dysentery and possibly the early stages of the heart disease that killed him in 1870; contemporaries noted he looked exhausted.

The Confederate army was missing Stonewall Jackson (killed at Chancellorsville), whose independent corps command Lee never adequately replaced.

Some subordinate failures (Ewell on July 1, Stuart's joyride) were not entirely Lee's fault.

The Army of Northern Virginia still managed to retreat in good order and fight another day.

Consensus Modern Verdict

Even strongly pro-Lee historians (e.g., Douglas Southall Freeman, Alan Nolan, Gary Gallagher) agree Gettysburg was Lee's worst battle:

Stephen Sears (Gettysburg, 2003): "Gettysburg was Robert E. Lee's poorest battle."

James McPherson: "Lee's defeat at Gettysburg marked the turning point when the Confederacy's hopes for independence began to fade."

Even the U.S. Army's official history calls July 3 "one of the most ill-advised attacks in American military history."

In short, while Lee remained a dangerous opponent after Gettysburg as Grant would soon find out, the battle revealed the limits of his aggressive style when facing a competent opponent on ground of the enemy's choosing. Gettysburg, more than any other engagement, cost the Confederacy its best chance for a negotiated peace and cemented Lee's reputation as a very good, but not infallible, general.

Casualties

Lee's losses were a catastrophic 43% of his army or 32,000 men. Union losses were similar but at a lower percentage. The north could replace its losses fairly easily, the south could not.

Battle of Gettysburg Casualty Comparison

Union						Confederate					
Strength	Killed	Wounded	Captured/ Missing	Total	%	Strength	Killed	Wounded	Captured/ Missing	Total	%
104,256	3,642	16,576	11,825	32,043	31%	75,000	NA	NA	NA	32,000	43%

Rating Lee's Performance at Gettysburg

Lee's generalship in the Gettysburg Campaign rates the lowest of all his Civil War battles/campaigns at an average 1.67 or *Bad/Poor*. Lee's strategic objective in conducting the campaign was to decisively defeat the Union army in a northern state to: 1) demoralize the northern public hoping they would demand an end to the war, and 2) convince Britain and France to recognize the Confederacy. Results at Gettysburg had the opposite effect—resulting in a Strategy score of Poor or 2.0.

Lee's tactics led to the defeat generating a poor score of 2.0. He continuously ordered his troops to assault fortified positions on his left, right, and finally center resulting in repulses each time. And, he failed to insist that his generals coordinate their attacks which allowed Meade to shift his forces to counter Lee's discordant attacks. Finally, Lee's futile attacks led him to lose almost half his army during the three-day battle. Losses Lee would never fully replace. It was the beginning of the end for the Confederacy.

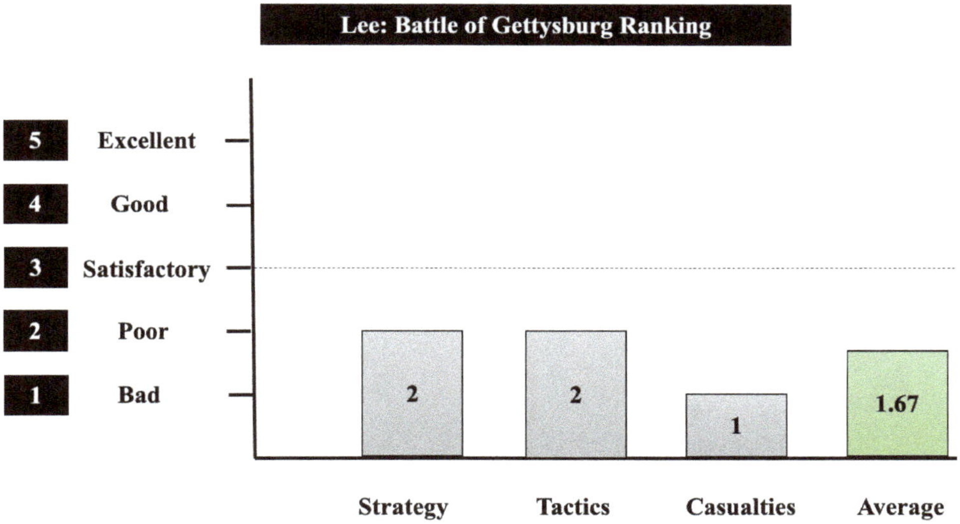

David W. Manly

Battle of Gettysburg—Day 1

Map by Hal Jespersen, www.cwmaps.com

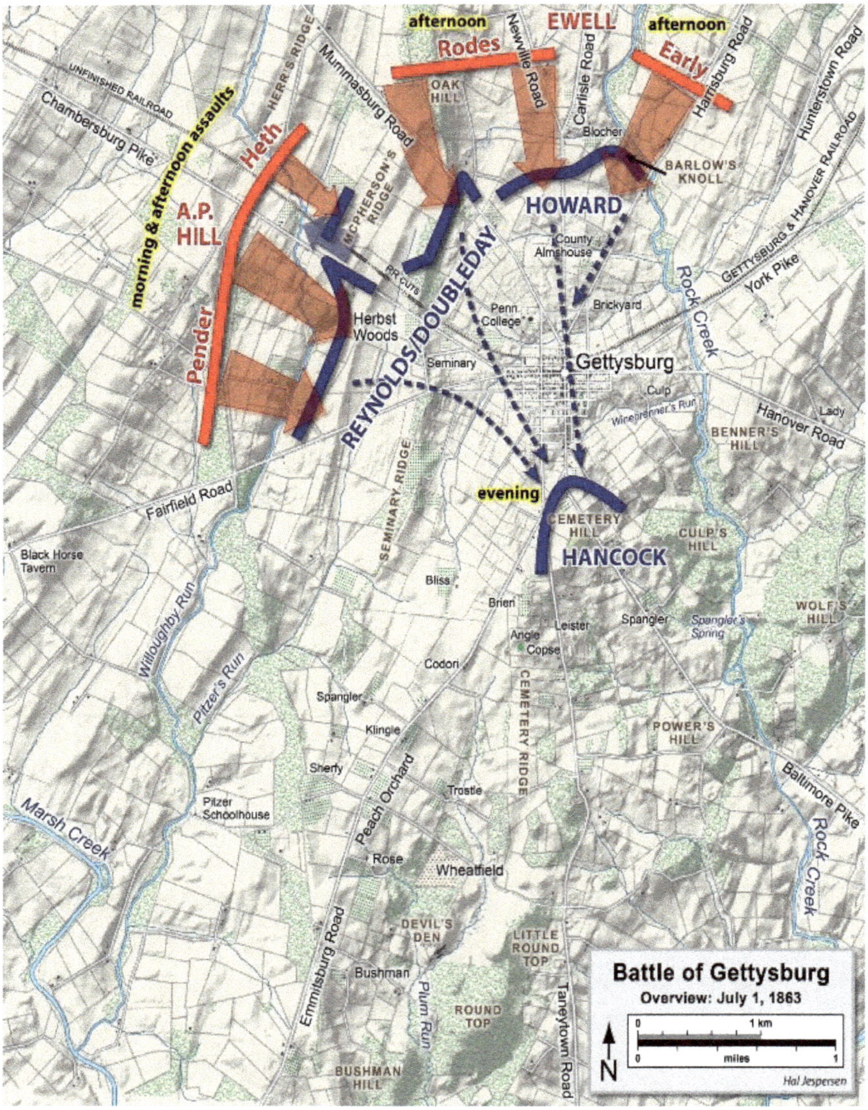

Battle of Gettysburg—Day 2

Map by Hal Jespersen, www.cwmaps.com

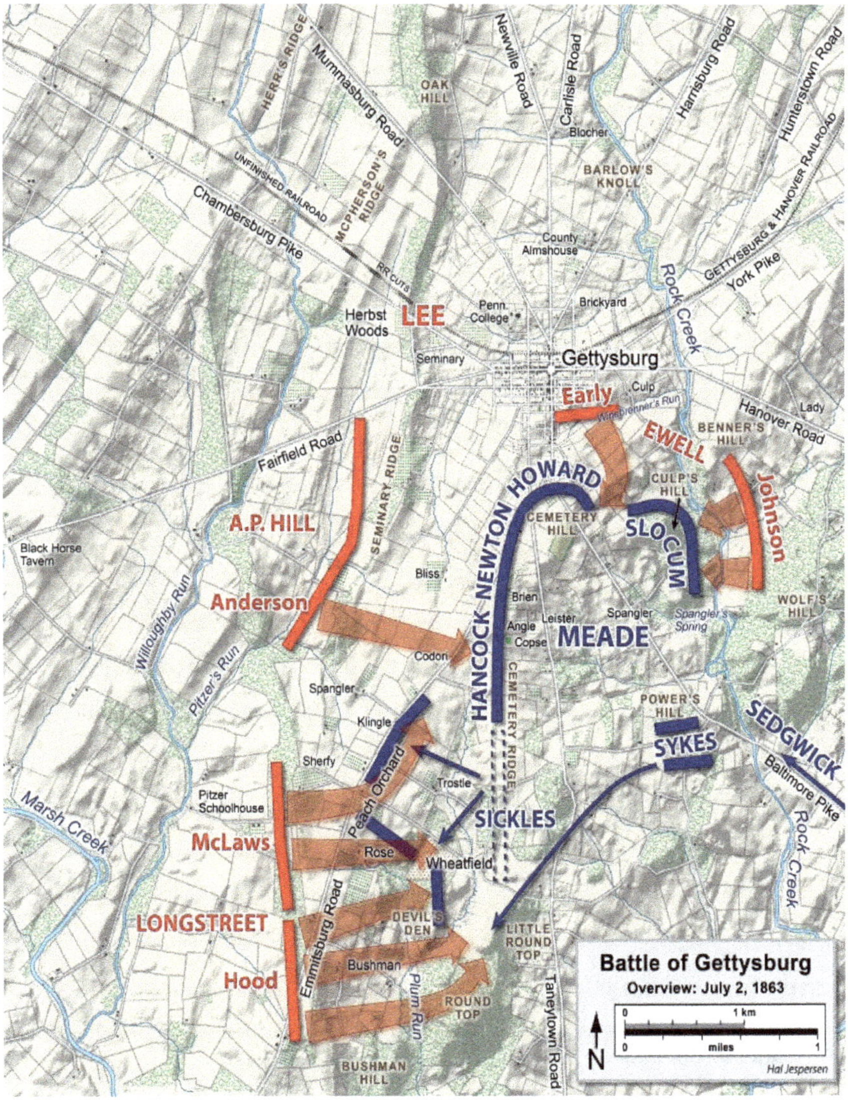

Battle of Gettysburg—Day 3

Map by Hal Jespersen, www.cwmaps.com

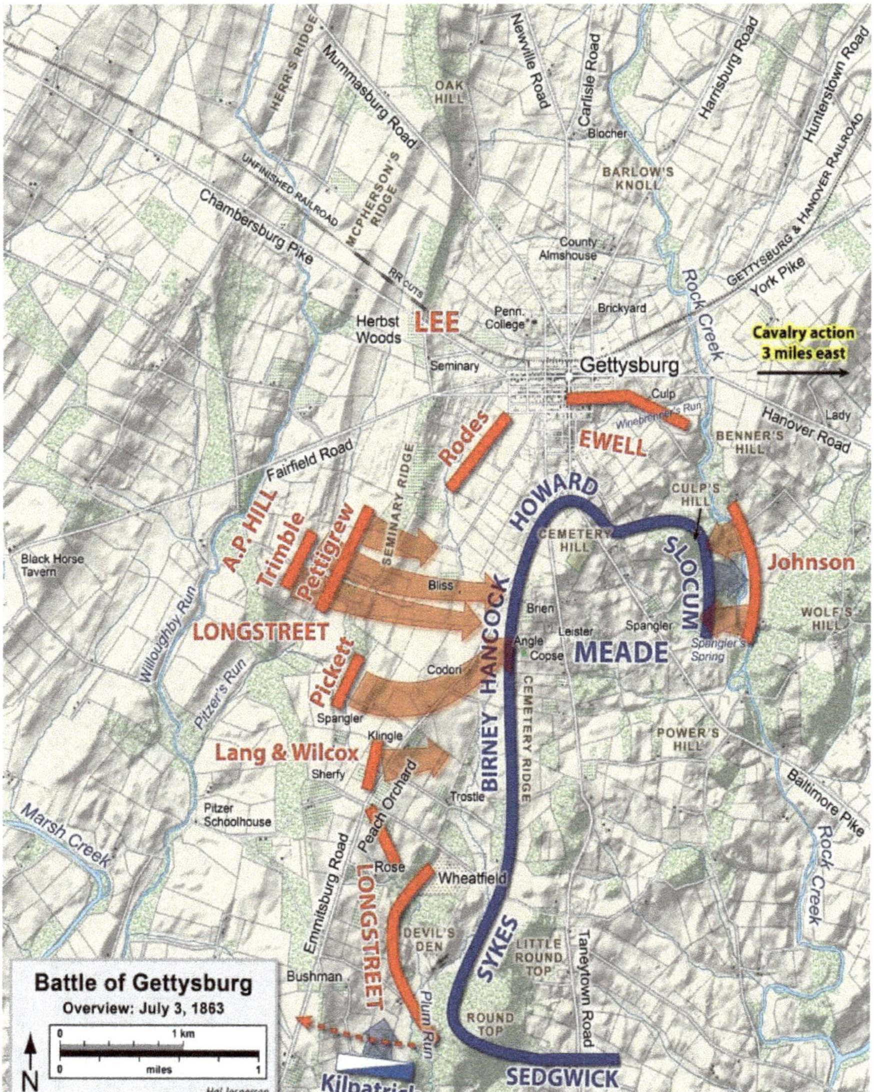

Grant: *Chattanooga/Missionary Ridge* (November, 1863)

Ulysses S. Grant's command during the Chattanooga Campaign stands as perhaps the best illustration of how the effect of Grant's confident, offensive demeanor translated to success on a battlefield. Combined with his prior success at Vicksburg, the campaign led Congress (championed by Lincoln) to appoint Grant leader of all Union armies.

Synopsis

The **Battle of Chattanooga** (November 23–25, 1863), often highlighted by the dramatic assault on **Missionary Ridge**, was a decisive Union victory that ended the Confederate siege of Chattanooga, Tennessee, and opened the gateway to the Deep South.

After the Union defeat at Chickamauga in September 1863, Maj. Gen. William S. Rosecrans's Army of the Cumberland retreated into Chattanooga, where it was besieged by Gen. Braxton Bragg's Army of Tennessee (about 44,000–48,900 men). The Confederates occupied the commanding heights of Lookout Mountain and Missionary Ridge, cutting off most Union supply lines and threatening starvation.

In mid-October, Grant arrived to take overall command. He replaced Rosecrans with Maj. Gen. George H. Thomas and opened a supply route dubbed the "Cracker Line". Reinforcements arrived: Maj. Gen. William T. Sherman's Army of the Tennessee and elements under Maj. Gen. Joseph Hooker.

The fighting began on November 23 with a Union demonstration at Orchard Knob, where Thomas's troops seized a forward Confederate position. On November 24, Hooker's forces captured Lookout Mountain in the famous "Battle Above the Clouds," driving Confederates off the heights in fog-shrouded combat. That same day, Sherman crossed the Tennessee River and attacked the northern end of Missionary Ridge (Tunnel Hill), but his assaults were repulsed by Lt. Gen. Patrick Cleburne's stout defense.

On November 25, Grant ordered Thomas's Army of the Cumberland to make a limited demonstration against the rifle pits at the base of Missionary Ridge to relieve pressure on Sherman. Instead, after capturing the lower trenches under heavy fire, the Union soldiers spontaneously continued uphill in one of the most remarkable assaults of the war. Defying orders to halt, they surged up the steep 500-foot slope, overran the poorly sited Confederate works on the crest, and shattered Bragg's center. Hooker simultaneously advanced from the south at Rossville Gap, rolling up the Confederate left.

Bragg's army broke and retreated in disorder toward Dalton, Georgia, abandoning artillery and supplies. A skillful rearguard action by Cleburne at Ringgold Gap slowed the Union pursuit.

Total casualties for the Chattanooga Campaign: Union approximately 5,800; Confederate roughly 6,700 (including about 4,000 captured).

The victory lifted the siege, secured Chattanooga as a vital Union base, severely damaged Confederate morale in the West, and set the stage for Sherman's Atlanta Campaign in 1864. It further solidified Grant's reputation, leading to his promotion to lieutenant general and command of all Union armies.

Key Events

1. **Relieving the Siege: The "Cracker Line" (October 1863):** Upon arriving in Chattanooga on October 23, Grant met with Rosecrans and determined the general was like a deer in headlights—unable to do what was needed to supply his troops and lift the siege. Grant immediately sat down at a desk and started writing orders to get things moving—without even taking is wet and muddy coat and boots off.

   Grant prioritized resupplying his 56,000-man Army of the Cumberland, which was down to rations for just days.

   He approved and oversaw Chief Engineer Maj. Gen. William F. Smith's audacious plan to reopen the Tennessee River supply route. Under cover of night on October 28–29, Union troops seized Brown's Ferry, constructed pontoon bridges, and established the "Cracker Line," delivering hardtack (crackers) and other essentials.

Grant's Role: His decisiveness stabilized morale and enabled offensive planning. As he later noted, Chattanooga was "the first battlefield I have ever seen where a plan could be followed and from one place the whole field be within one view," allowing him to direct operations from a hill in front of Missionary Ridge called Orchard Knob.

2. **Battle of Lookout Mountain (November 24, 1863):** With reinforcements from Maj. Gen. Joseph Hooker (15,000 troops) and Sherman (20,000), Grant launched a three-day offensive. Grant ordered Hooker to assault Lookout Mountain, a fog-shrouded 1,200-foot peak southwest of Chattanooga held by about 1,800 Confederates. Despite steep terrain and initial resistance, Hooker's forces routed the enemy in the "Battle Above the Clouds," forcing their retreat by afternoon.

   Grant's Role: Observing from Orchard Knob, Grant calmly adjusted plans amid success, using the victory to threaten Bragg's left flank. This diversionary action succeeded beyond expectations, clearing the way for the main assault.

3. **Battle of Missionary Ridge (November 25, 1863):** Bragg's 44,000 Confederates were entrenched along the 5-mile Missionary Ridge, a 300-foot escarpment east of Chattanooga overlooking the city.

   Grant's plan: Sherman attacks the northern end (Tunnel Hill) to "roll up" the line; Hooker strikes the southern end from Lookout; Maj. Gen. George H. Thomas's Army of the Cumberland demonstrates in the center to pin Bragg.

   Sherman faced delays due to poor terrain intelligence, achieving only partial success. Hooker's advance stalled in Chattanooga Valley. In the center, Thomas's men seized Confederate rifle pits at the base but, defying Grant's orders to halt and consolidate, spontaneously charged uphill under heavy fire, overwhelming the thinly held crest in a stunning rout.

   Grant's Role: Astonished by the improvised assault (which he initially viewed skeptically), Grant adapted seamlessly, committing reserves to exploit the breakthrough. His forces captured 4,000+

prisoners and forced Bragg's retreat into Georgia, with Union casualties at 5,824 vs. Confederate 6,667.

Overall Assessment of Grant's Performance

Grant's Chattanooga victory showcased his hallmarks: unflinching resolve, logistical acumen, and ability to integrate disparate armies (Cumberland, Tennessee, Potomac detachments) into a cohesive force. Critics note the center charge's disobedience, but Grant praised Thomas's troops, crediting their momentum and Bragg's errors (e.g., detaching Lt. Gen. James Longstreet to Knoxville). The campaign boosted Grant's reputation, leading President Lincoln to promote him to lieutenant general in 1864 as commander of all Union armies.

Grant's Performance at Chattanooga/Missionary Ridge		
Aspect	**Grant's Approach**	**Result/Impact**
Logistics	Established "Cracker Line" at Brown's ferry allowing supplies to enter town	Ended starvation of men and horses; enabled his 40,000+ men to rally and fight
Coordination	Multi-Army plan: Sherman (north), Hooker (south), and Thomas (center)	Fully lifted the siege, Bragg retreated 50+ miles south
Adaptability	Adapted to fog, delays, and the spontaneous charge of his troops at the base of Missionary Ridge	Secured the crucial Chattanooga rail hub; allowing for invasion of the Confederate deep south
Casualties	Union: about 5,700; Confederate about 6,700	Decisive Union edge; huge morale booster for his troops; completely turned around the situation in the Union's favor

This performance solidified Grant as the Union's indispensable leader, transforming Chattanooga from a trap into a launchpad for total war.

Casualties

Despite being the attacking force, Grant's casualties were lower both in the absolute and percentage wise compared to Bragg's. For a relatively light loss, Grant freed the key city of Chattanooga from a Confederate siege, re-invigorated the Union Army in Tennessee, and forced the Confederates to retreat into Georgia leading to Sherman's eventual "march to the sea".

Battle of Chattanooga/Missionary Ridge Casualty Comparison

Union						Confederate					
Strength	Killed	Wounded	Captured/ Missing	Total	%	Strength	Killed	Wounded	Captured/ Missing	Total	%
56,359	753	4,722	349	**5,824**	**10%**	44,010	361	2,160	4,146	**6,667**	**15%**

Rating Grant's Performance at Missionary Ridge

Grant's strategic objectives at Chattanooga were to free the Army of the Cumberland from the siege which threatened their annihilation, and drive the Confederates out of the border state of Tennessee. Grant's victories at Lookout Mountain and Missionary Ridge accomplished his objectives for a 5.0 score on Strategy. Grant's tactics rate a 4.0 as they were generally Good and would be Excellent if not for Sherman's failure to advance faster due to poor understanding of the terrain, and the fact Thomas's troops stormed Missionary Ridge without orders, completely at their own initiative. Grant's total casualties were 10% of his force slightly less than Bragg's. Overall score: 4.33 or *Good* bordering on *Excellent.*

Grant: Battles of Chattanooga/Missionary Ridge Ranking

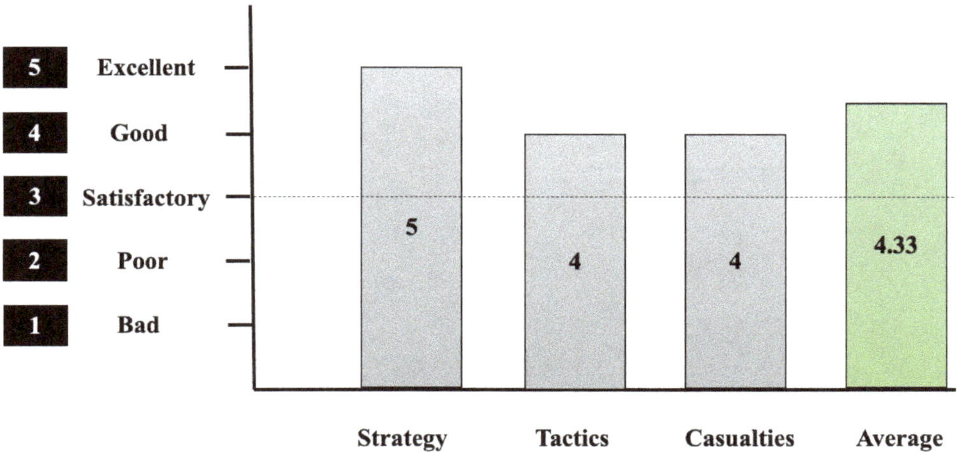

Battles of Chattanooga and Missionary Ridge

Map by Hal Jespersen, www.cwmaps.com

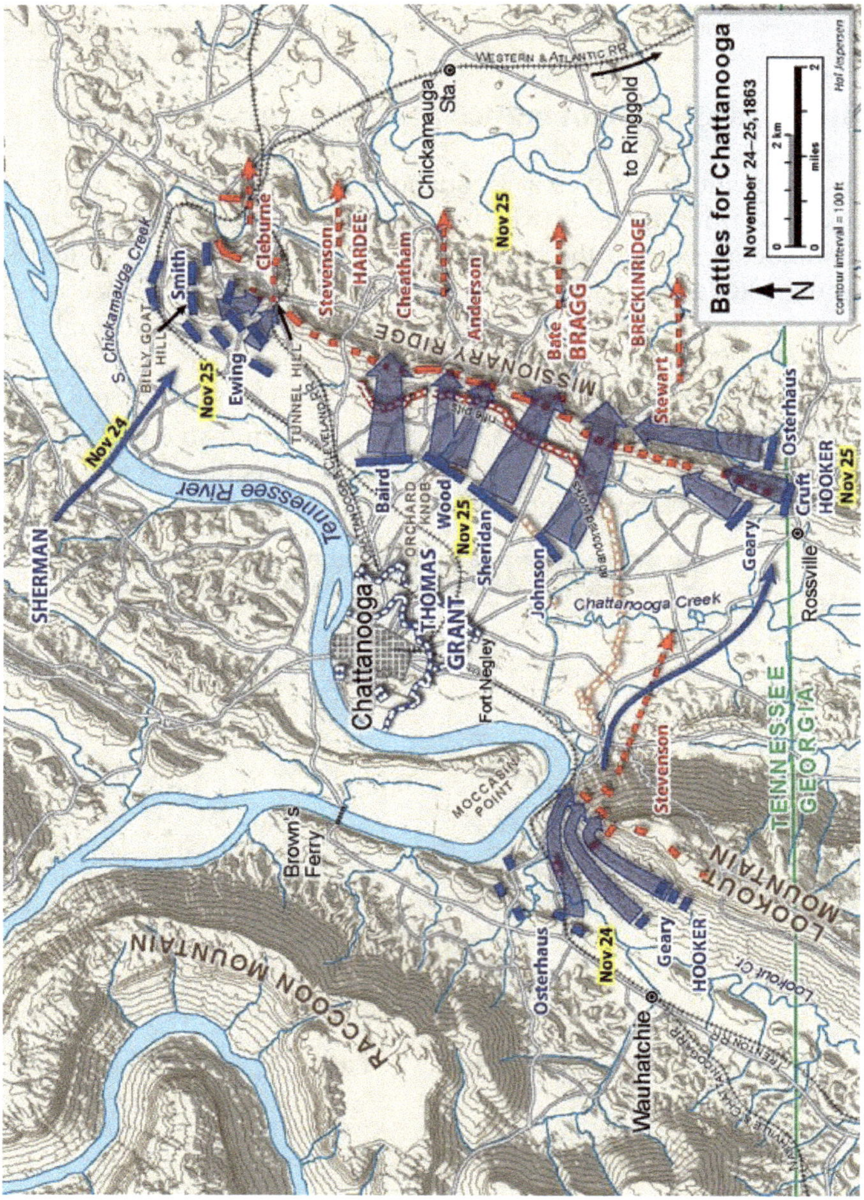

David W. Manly

Lee and Grant: *The Overland Campaign* (May/June, 1864)

The Overland Campaign was Ulysses S. Grant's relentless 40-day offensive in Virginia designed to destroy Lee's Army of Northern Virginia and end the war. While Lee's tactics thwarted Grant battle after battle, Grant successfully maintained the initiative and achieved his strategic goal of decimating the Confederate army leading to eventual defeat a year later.

Synopsis

The **Overland Campaign** (May 4–June 24, 1864), was Grant's aggressive offensive in Virginia designed to end the war. Newly promoted to general-in-chief of all Union armies by an act of Congress, Grant accompanied Maj. Gen. George G. Meade's Army of the Potomac (approximately 120,000 men) in a relentless drive to destroy Gen. Robert E. Lee's Army of Northern Virginia (about 65,000 men) or force it into a defensive siege, rather than merely capturing Richmond.

The campaign opened on May 4 when Union forces crossed the Rapidan River. On May 5–7, the armies clashed in the dense, tangled **Battle of the Wilderness**. Lee attacked aggressively despite being outnumbered, and the fighting devolved into chaotic, small-unit combats amid thick underbrush that caught fire, trapping wounded men. Both sides suffered heavy losses but Grant refused to retreat. Instead of withdrawing like previous Union commanders, he ordered a nighttime march southeast toward Spotsylvania Court House on May 7–8, maintaining pressure on Lee.

At **Spotsylvania Court House** (May 8–21), Lee arrived first and entrenched. The battle featured nearly two weeks of brutal fighting, highlighted by the May 12 assault on the "Bloody Angle" or "Mule Shoe" salient. Union troops under Maj. Gen. Winfield Scott Hancock broke through, leading to nearly 24 hours of savage hand-to-hand combat—the longest continuous infantry fight of the war. Grant again sidestepped after failing to achieve a decisive breakthrough.

Subsequent maneuvers included the **Battle of North Anna** (May 23–26), a brief but sharp engagement where Lee attempted to trap parts of Grant's army, and cavalry actions such as the death of Confederate cavalry leader J.E.B. Stuart at Yellow Tavern on May 11.

The campaign culminated at **Cold Harbor** (May 31–June 12), near Richmond. On June 3, Grant ordered a massive frontal assault against well-fortified Confederate lines. The attack was a disaster. Grant later expressed deep regret over the assault.

On June 12–18, Grant executed a brilliant crossing of the James River and shifted operations toward Petersburg, the vital rail hub south of Richmond, effectively ending the Overland phase and beginning the ten-month Siege of Petersburg.

Tactically, the campaign featured no single decisive victory for either side, and Grant endured heavy criticism in the North for the butcher's bill. Strategically, however, it was a major Union success. Grant's policy of continuous contact and maneuver prevented Lee from regaining the initiative, inflicted irreplaceable attrition on the Confederacy, and pinned Lee's army into static defenses around Richmond and Petersburg. This allowed coordinated Union advances elsewhere, including Sherman's campaign in Georgia. The Overland Campaign marked the beginning of the war's final phase of attrition that ultimately led to Confederate defeat.

Lincoln Commissions Grant as Lieutenant General at the White House

Key Features

Opposing Commanders: Lt. Gen. Ulysses S. Grant & Maj. Gen. George G. Meade (Army of the Potomac) vs. Gen. Robert E. Lee (Army of Northern Virginia)

Union strength: roughly 118,000 at the start

Confederate strength: roughly 66,000 at the start

Casualties (approximate, May 4–June 12):

- Union: ~68,000 killed, wounded, missing, captured
- Confederate: ~36,000 killed, wounded, missing, captured

Nicknamed the "Forty Days" or "Grant's Hammering Campaign"

Major Battles (in sequence)

1. **Battle of the Wilderness (May 5–7, 1864):** Dense woods near Chancellorsville; brutal, confused fighting. Union losses ~18,000, Confederate ~11,000. Instead of retreating north as previous Union generals had done, Grant sidled south toward Spotsylvania.

2. **Battle of Spotsylvania Court House (May 8–21, 1864):** Lee entrenched quickly. Highlight: The Bloody Angle (May 12), 20+ hours of hand-to-hand combat in the rain over a salient in Lee's line. Union ~18,000 casualties, Confederate ~12–13,000. Grant again slid southeast.

3. **Battle of North Anna (May 23–26, 1864):** Lee formed an inverted "V" trap across the river, but illness (severe diarrhea) kept him from springing it aggressively. Minor fighting; Grant again maneuvered around.

4. **Battle of Toponomy Creek / Bethesda Church (May 28–31, 1864):** Skirmishing as Grant continued moving southeast.

5. **Battle of Cold Harbor (May 31 – June 12, 1864**: One of the war's most lopsided attacks. On June 3, ~7,000 Union soldiers fell in less than an hour assaulting fortified lines. Total Union losses

~13,000, Confederate ~5,000. Grant later wrote in his memoirs that he "always regretted" the last assault at Cold Harbor.

6. **Assault on Petersburg (June 15-18, 1864):** Union forces attack Petersburg—a key rail center and gateway to Richmond which is 23 miles north. Union forces narrowly miss capturing the city due to poor coordination and timely Confederate reinforcements leading to a prolonged siege.

Union and Confederate Troops Clash in the Wilderness

Outcome & Strategic Result

After Cold Harbor, instead of retreating or stopping, Grant executed a brilliant strategic maneuver: he disengaged, crossed the James River on a 2,200-foot pontoon bridge (the longest in military history to that point), and marched on Petersburg, the key rail hub supplying Richmond. Lee was caught off-guard and barely raced his army south in time.

The campaign ended with the beginning of the Siege of Petersburg (June 9, 1864 – April 2, 1865), which ultimately doomed the Confederacy.

Why It Matters

- **No Retreat!** For the first time a Union commander refused to retreat after heavy casualties in Virginia.
- **War of Attrition**: The campaign shifted the war in the East from maneuver-and-politics to a modern war of attrition.

- **Beginning of the End**: The irreplaceable Confederate losses (especially officers) began the irreversible erosion of the Army of Northern Virginia.
- **Election Impact**: Public and political shock in the North over casualty lists nearly cost Lincoln the 1864 election (until Sherman took Atlanta in September).

In short, the Overland Campaign was the bloodiest six weeks of the entire war, but it put the Confederacy on an irreversible path to defeat by tying Lee down in trenches south of Richmond and bleeding his army white.

Grant's performance in the Overland Campaign was strategically excellent and ultimately decisive**, but** tactically flawed and extremely costly. Historians almost universally rank it as one of the most important Union campaigns of the war, while acknowledging its brutal price and several operational mistakes.

What Grant Got Right (Strategic Mastery)

The Overland Campaign: What Grant Got Right	
Aspect	**Why It Was Brilliant**
Relentless Pressure	Refused to retreat after the Wilderness, Spotsylvania or Cold Harbor; broke the psychological pattern of previous Union forays into Northern Virginia
War of Attrition	Accepted that the Confederacy could not replace losses; deliberately traded 1.6 to 1.7 Union casualties for every Confederate to bleed Lee dry
Coordination with other Theaters	Kept Butler, Sigel, and later Hunter moving in secondary offensives to prevent Lee from securing fresh troops. After June redirected the campaign under Sheridan to clear the Shenandoah Valley
Final Maneuver to Petersburg	The undetected crossing of the James river on June 12-15 with over 100,000 of his army completely surprised Lee. It is the largest river crossing of any army in the war
Kept the Initiative	For 40 straight days Lee was reacting; he was never able to regain the initiative after the Wilderness

Result: Lee's army was pinned south of Richmond in trenches by mid-June 1864 and never conducted another strategic offensive. The Confederacy was doomed once Petersburg was besieged.

What Grant Got Wrong

The Overland Campaign: What Grant Got Wrong	
Mistake/Problem	**Details**
Poor Battlefield Control	Grant delegated battlefield tactical command to Meade; but the Grant-Meade relationship was often tense; leading to often vague or countermanded orders to subordinates
Over-Aggressive Frontal Assaults	Repeated direct frontal attacks on entrenched Confederate positions (Spotsylvania "Bloody Angle", Cold Harbor (June 3) produced horrific casualties for negligent gain
Cold Harbor (June 3)	The single worst tactical decision of Grant's entire military career; produced 7,000+ casualties in less than an hour—for no gain.
Under-Use of Calvary Early	Did not initially employ Sheridan aggressively; only after Cold Harbor was Sheridan given full reign resulting in victories at Trevilian Station, Yellow Tavern and later in the Valley campaign.
High Desertion & Morale Dip	By June, 1864 the Army of the Potomac was exhausted; straggling and desertions spiked; and Northern public opinion turned against "Grant the Butcher".

Casualty Ratio Reality Check

Raw numbers: ~68,000 Union vs ~36,000 Confederate. A ratio of 1.88: 1.

But the Confederacy could not replace losses, while the Union could (and did). Grant knew this and accepted the math.

Modern historians (e.g., Gordon Rhea, Mark Grimsley) argue the ratio was the best the Union could realistically achieve against Lee in 1864.

Contemporary & Historical Verdicts

Consensus About the Overland Campaign	
Source	**Verdict**
Lincoln (June, 1864)	*"Grant is remorselessly driving Lee into Richmond". "I can't spare this man, he fights"*
Northern Newspapers (Summer, 1864)	Called him "Butcher Grant"; a huge political liability until Sherman took Atlanta in September
Robert E. Lee (postwar)	Privately admitted the "campaign wore us out in the end"
Modern Consensus (Bonekemper, Rhea, Simpson, others)	Strategically one of the three most important campaigns of the war (along with Vicksburg and Atlanta). Tactically wasteful but strategically decisive

Bottom Line

Grant's Overland Campaign performance achieved exactly what he set out to do: neutralize Lee's army and set up the final stage of conquest. Tactically Grant paid an unnecessarily high price in several battles, especially Cold Harbor. But overall the campaign broke the Army of Northern Virginia's ability to maneuver and placed the Confederacy in a death grip from which it never escaped.

It was ugly, it was bloody, and at times it was clumsy — but it worked, and it shortened the war. That is why many historians rank the Overland Campaign as Grant's second-greatest achievement after Vicksburg.

Lee and the Overland Campaign

The Overland Campaign (May–June 1864) was Robert E. Lee's most difficult and perhaps controversial campaign of the American Civil War. Facing Ulysses S. Grant for the first time, Lee commanded the Army of Northern Virginia against a Union force that was larger, better supplied, and—crucially—willing to accept heavy casualties to achieve strategic ends.

Historians remain deeply divided on how well Lee performed during the Overland Campaign. Here is a balanced assessment broken down by key criteria:

Lee's Strategic Performance: Mixed, leaning toward failure

1. **Lee correctly divined that Grant would march through the Wilderness** area and fought the opening battle (Wilderness, 5–7 May) exactly where he wanted—dense terrain that neutralized Union artillery and numbers.

2. **However, Lee never developed a coherent operational plan beyond reacting to Grant's moves**. Grant's strategy was simple but brutal: keep advancing south and force Lee into a battle of attrition. Lee could not stop the southward slide (Wilderness → Spotsylvania → North Anna → Cold Harbor → Petersburg).

3. **By the end of six weeks, Lee had been maneuvered out of central Virginia and bottled up in the trenches around Richmond and Petersburg**—exactly the outcome he had always said he must avoid.

Tactical Performance: Superb at times, catastrophic at others

Excellent:

- **Wilderness** (5–7 May): Classic Lee—aggressive counterattacks (Longstreet's flank attack on 6 May) inflicted ~18,000 Union casualties vs ~11,000 Confederate.

- **Spotsylvania** (8–21 May): The Bloody Angle (22 hours of hand-to-hand fighting on 12 May) is one of the war's most ferocious engagements; Lee's men held the vital "Mule Shoe" long enough to build a new line.

Disastrous:

- **Allowing Grant to steal a march and cross the James River unopposed**. Grant successfully pulled his troops off the line without Lee finding out for about 24 hours. Grant's men-built pontoon bridges across the James and crossed with no opposition. His army reached Petersburg before Confederate reinforcements but Grant's subordinates got cold feet when they a skeleton crew of rebels in trenches and they failed to attack. If Grant had been on-hand (he was supervising the crossing) Petersburg would have fallen in 1864 instead of a year later in 1865. The war might have been over then.

- **Lee was unusually passive in command style** during this campaign. He delegated too much to corps commanders (especially Ewell and A.P. Hill) who were in poor health or simply not up to the task.
- **Failure to coordinate the army on several occasions** (e.g., almost losing the Mule Shoe at Spotsylvania because Lee pulled artillery away prematurely; the near-disaster on the North Anna when Lee was bedridden and issued almost no orders).
- **Casualties were irreplaceable**: ~36,000 Confederate losses (killed, wounded, missing) in six weeks—roughly 50–55% of the army Lee started with in May. This was proportionally worse than Grant's losses.

Overall Historical Verdict

Modern scholarship (Gary Gallagher, Gordon Rhea, Mark Grimsley, etc.) tends to rate Lee's Overland Campaign performance as one of his weakest:

1. **Grant's strategy trumped Lee's tactics**: Lee won every major tactical engagement on points (except perhaps North Anna, a draw), yet lost the campaign decisively.

2. **Unacceptable casualties**: Lee inflicted ~68,000 Union casualties but at the unsustainable cost of ~36,000 of his own—ratio about 1.88:1 in his favor, far below the 3:1 or 4:1 he needed given the disparity in total resources.

3. **War on Grant's Terms**: Most importantly, Grant achieved his strategic goal: Lee was forced into siege warfare, which the Confederacy could not win.

Consensus modern rating: Lee was a tactical genius who was strategically outgeneraled and bled his army white

Lost Cause/Lee-centric historians (e.g., Douglas Southall Freeman earlier in the 20th century) gave him an A or A+ because "he never lost a battle." Contemporary historians almost uniformly reject that interpretation for the Overland Campaign.

Short verdict: Lee fought with his usual brilliance in set-piece battles but was completely outmaneuvered at the operational and strategic levels, and the Army of Northern Virginia never recovered from the losses it suffered under his command in May–June 1864.

Casualties

The Overland Campaign was a slaughter for both sides. Both Lee and Grant lost over half their strength they had at the Battle of the Wilderness. However, Grant's total losses of 64,000+ were almost double Lee's losses of 34,000+.

In his *Memoirs* Grant lamented the losses but argued: 1) an attacking army almost always suffers more casualties than the defender, 2) Lee had the advantage of interior lines and familiar geography, and 3) the losses were justified as they shortened the war and thus overall saved lives as more soldiers died from non-combat reasons than combat (dysentery, typhoid fever, malaria). Bottom-line: the north could replace their losses the south could not.

Overland Campaign Casualty Comparison

	Union						Confederate					
	Strength	Killed	Wounded	Captured/ Missing	Total	%	Strength	Killed	Wounded	Captured/ Missing	Total	%
Wilderness	118,700	2,246	12,037	3,383	17,666	15%	66,140	1,477	7,866	1,690	11,033	17%
Spotsylvania	115,400	2,725	13,416	2,258	18,399	16%	63,000	1,515	5,414	5,758	12,687	20%
North Anna	100,000	591	2,734	661	3,986	4%	53,000	124	704	724	1,552	3%
Cold Harbor	117,000	1,845	9,077	1,816	12,738	12%	62,000	788	3,376	1,123	5,287	9%
Petersburg	62,000	1,688	8,513	1,185	11,386	18%	38,000	200	2,900	900	4,000	11%
Total		9,095	45,777	9,303	64,175	54%		4,104	20,260	10,195	34,559	52%

David W. Manly

Rating Grant and Lee's Performances in the Overland Campaign

Grant's strategic objectives were two-fold: 1) decimate if not destroy the Lee's army, and 2) threaten the capture of Richmond the capital of the Confederacy. Grant largely achieved these objectives as Lee's fighting forces suffered irretrievable losses of over 34,000 men, and the encirclement and siege of Petersburg more or less ensured the eventual capture of Richmond and the end of the war. For Strategy we rate Grant a 5 or Excellent.

Tactically Grant's efforts were mixed. On the plus side he continually maintained the initiative forcing Lee to react to his moves. Additionally, Grant flawlessly moved his entire army across the James River in the face of the enemy resulting in a complete surprise to Lee. On the other hand, Grant's continued assault of entrenched positions at Spotsylvania and Cold Harbor led to extreme losses with little or no gain. Grant's tactics in these two battles showed little imagination or creativity. Tactics score overall 2 or Poor. Casualty score is a 2 or Poor due to the extraordinary losses from his frontal assaults. Overall: 3.0 or *Satisfactory.*

Lee's strategic objective was to repulse Grant's army from Virginia as he had with McLellan, Hooker, and Burnside before. He failed in his objective as Grant was a force Lee had never seen before. Strategy score of 2 or Poor. Tactically Lee deserves a higher score of 4 or Good for a couple of reasons. First, Lee was partially successful in blunting Grant by marching forcibly to meet him in the Wilderness, and secondly by stealing the march to the important crossroads of Spotsylvania arriving just ahead of Union troops. Additionally, Lee's troops successfully resisted Grant at the Mule Shoe and at Cold Harbor inflicting terrible losses.

Lee's Tactics score is only Good and not Excellent as he was never able to retrieve the initiative from Grant, and he allowed Grant to move his entire army across the James while directly opposite his lines. From a Casualty standpoint, Lee's performance is 2 or Poor as he lost approximately half his army or over 34,000 total. Total score of 2.67 or *Poor/Satisfactory.*

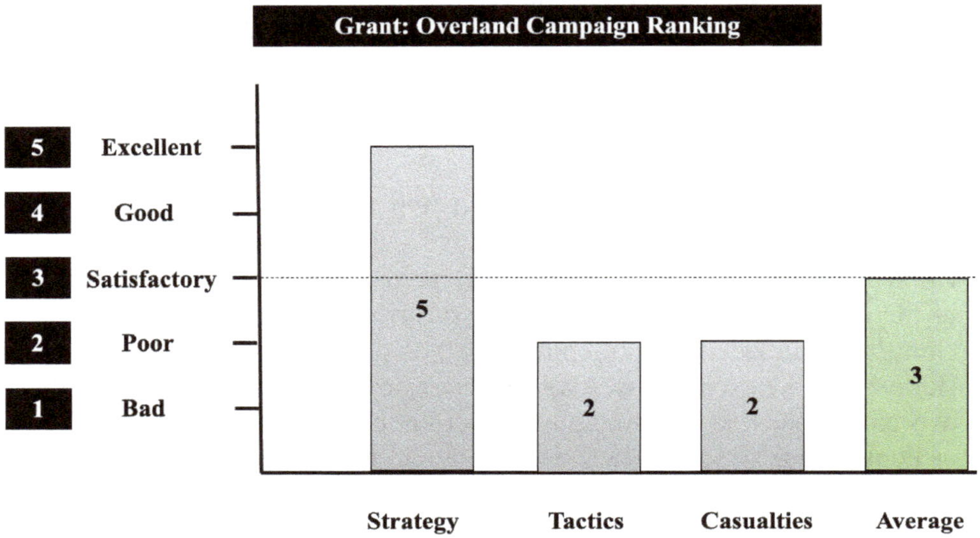

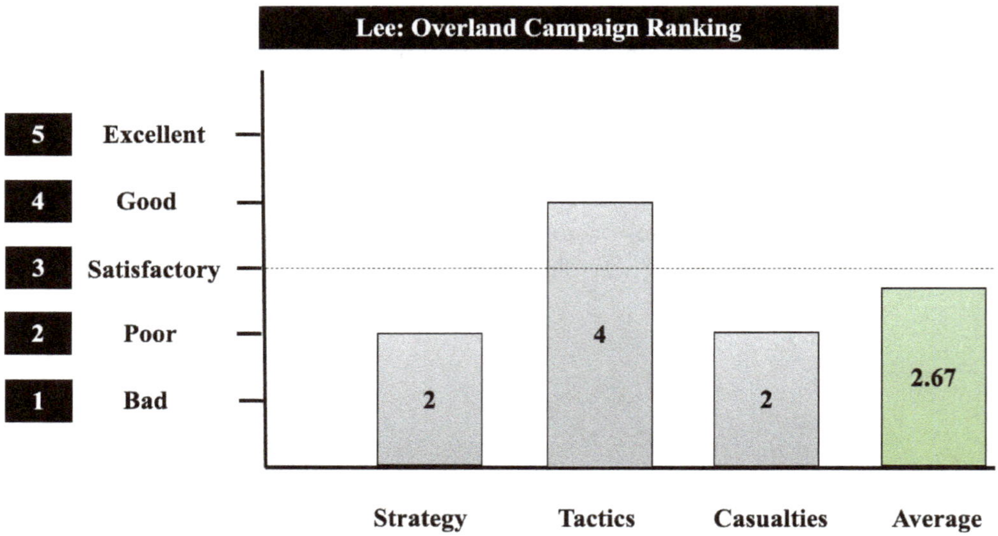

The Overland Campaign

Map by Hal Jespersen, www.cwmaps.com

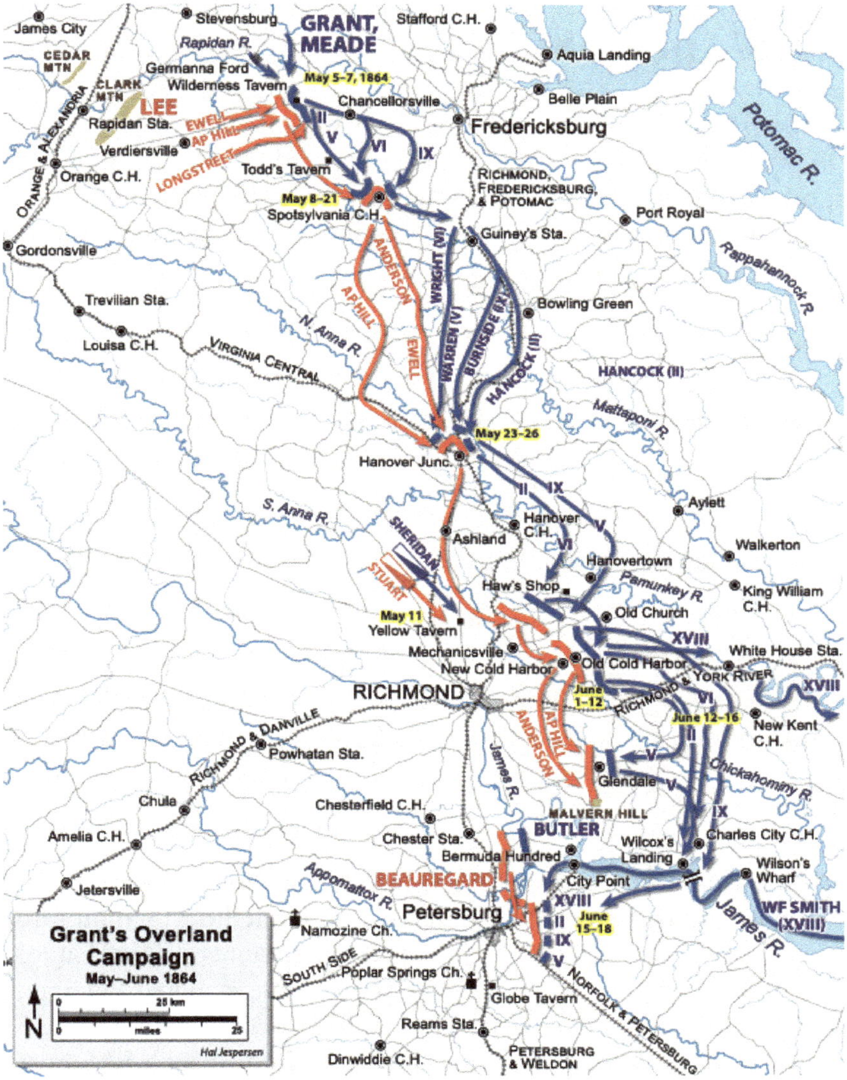

Lee and Grant: *Appomattox Campaign* (March/April, 1865)

The Appomattox Campaign was the final military operation of the Civil War in Virginia. The campaign is widely regarded as a textbook example of Grant's strategic and operational mastery, and simultaneously one of the clearest demonstrations of Lee's deteriorating options and the limits of even his considerable tactical skill.

Synopsis

The **Appomattox Campaign** (March 29–April 9, 1865) was the final offensive of the Civil War in Virginia. It began with Grant's Union forces breaking the ten-month Siege of Petersburg and ended with Lee's surrender of the Army of Northern Virginia at Appomattox Court House, effectively concluding major combat in the Eastern Theater.

After months of attrition around Petersburg and Richmond, Grant launched a flanking movement southwest of Petersburg on March 29 to cut Lee's remaining supply lines. The decisive blow came at the Battle of Five Forks on April 1, where Maj. Gen. Philip Sheridan's cavalry and V Corps crushed Confederate forces under Maj. Gen. George Pickett, opening the Southside Railroad. Lee lost roughly 2,950 men there.

On April 2, Grant ordered a general assault. The **Third Battle of Petersburg (Breakthrough)** shattered Confederate lines; Union forces overran trenches, inflicting heavy losses (Confederates suffered at least 5,000 casualties, many captured). That night, Lee evacuated Petersburg and Richmond, beginning a desperate westward retreat with about 35,000 starving, exhausted troops, hoping to link with Gen. Joseph E. Johnston in North Carolina.

The retreat turned into a running fight. On April 6 at **Sayler's Creek** (near Farmville), Union forces under Sheridan, Maj. Gen. Horatio Wright, and Andrew Humphreys trapped large portions of Lee's army. In three separate engagements, Confederates lost over 7,500 men and eight generals—one of the largest single-day surrenders of the war. Lee reportedly exclaimed, "My God! Has the army been dissolved?"

Lee reached Farmville on April 7 but could not resupply adequately. Grant sent a letter urging surrender. Lee continued west but was blocked at Appomattox Station on April 8, where Union cavalry captured critical supply trains.

On the morning of April 9, Lee attempted one last breakout west of Appomattox Court House with John B. Gordon's corps and Fitzhugh Lee's cavalry. They initially pushed back Sheridan's horsemen, but overwhelming Union infantry (including the Army of the James) appeared on both flanks. Realizing further resistance was futile, Lee met Grant at the McLean House around 1:30 p.m.

Grant offered generous terms: officers and men could return home with their horses and sidearms, and no one would be prosecuted for treason. Lee accepted, and approximately 28,000 Confederates were paroled. Formal surrender ceremonies occurred over the following days.

Union casualties for the campaign were about 5,300 total; Confederate losses of the entire Army of Northern Virginia exceeded 45,000. The surrender triggered the collapse of other Confederate armies and brought the war to a close within weeks.

The Appomattox Campaign demonstrated Grant's relentless pressure and Lee's skillful but ultimately hopeless retreat. It remains a symbol of reconciliation, highlighted by the dignified and magnanimous behavior of both commanders.

Union Entrenchments at Petersburg

Ulysses S. Grant's Performance

Grant's conduct of the campaign is generally considered a model of modern pursuit warfare.

Key strengths demonstrated:

1. **Strategic vision**: After the failure to destroy Lee at Petersburg/Hatcher's Run in late March, Grant immediately launched the "left hook" via Dinwiddie Court House and Five Forks (April 1) that unhinged Lee's right flank and forced the evacuation of both Petersburg and Richmond.

2. **Coordination of multiple armies**: Grant kept the Army of the Potomac (Meade), Army of the James (Ord), and Sheridan's cavalry operating in close concert. Sheridan's independent cavalry-infantry force (with V Corps and portions of II and VI Corps) became the decisive maneuver element that repeatedly got west of Lee and blocked his escape.

3. **Relentless pursuit**: From April 2–8 Grant drove his men mercilessly (marches of 25–30 miles per day in mud and rain) while maintaining supply lines and communications. He refused to let Lee break contact or reach a defensible position/supply base.

4. **Operational foresight**: Grant correctly predicted Lee would try to move south-west along the Richmond & Danville Railroad toward Danville or Lynchburg to link with Johnston. He positioned forces (Sheridan, Ord, and Griffin) to intercept at every possible point (Jetersville, Farmville, Appomattox Station, Appomattox Court House).

5. **Psychological pressure**: Grant sent letters to Lee (April 7, 8, 9) that were firm but courteous, offering generous terms and emphasizing the hopelessness of further resistance without being vindictive.

Result: Lee was brought to bay in nine days with relatively light Union casualties (~5,000 total for the campaign) while inflicting disproportionate loss on the Confederates and capturing or scattering the entire Army of Northern Virginia. This was Grant's third capture of a rebel army after Fort Donelson, and Vicksburg.

Modern military historians (e.g., Gordon Rhea, William Marvel, Earl Hess) almost unanimously rate Grant's handling of the campaign as brilliant.

The Fall of Richmond

Robert E. Lee's Performance

Lee's options were extremely constrained by April 1865, but his decisions still receive a mixed evaluation.

What Lee did well:

1. **Rapid and orderly evacuation of the Petersburg–Richmond lines** on the night of April 2–3, saving the bulk of his army from immediate capture.

2. **Attempted to keep the army concentrated** and moving toward supposed supply depots (Amelia Court House, Danville).

3. **Tactical success in small rear-guard actions** (e.g., Sutherland's Station, Namozine Church, Sailor's Creek—though the latter was a disaster).

A few criticisms:

1. **Miscalculation about supplies**: Lee expected 350,000 rations to be waiting at Amelia Court House (April 4–5). None were there because the railroad had already been cut and orders were garbled.

The resulting 24–48-hour delay while foraging was fatal; it allowed Grant's forces to leapfrog ahead and block the line of retreat.

2. **Failure to break contact cleanly**: After Amelia, Lee continued marching west instead of attempting a bold breakout north across the Appomattox River toward Lynchburg earlier, or dispersing into the mountains for guerrilla warfare (which he explicitly rejected).

3. **Lack of consolidation**: At Sayler's Creek (April 6): Roughly one-quarter to one-third of his remaining army (≈7,700 men) was lost in a running fight when Ewell's and Anderson's corps were cut off and destroyed or captured. Many historians see this as the point where organized Confederate resistance effectively ended.

4. **Underestimation of Union cavalry**: Sheridan's troopers, now armed with Spencer repeating rifles and operating semi-independently, repeatedly outmaneuvered and outfought Stuart's successors (Fitzhugh Lee and Rosser).

Assessment: Lee fought with his usual tenacity and maintained remarkable cohesion until the very end, but the campaign revealed the limits of inspiration and tactical brilliance when facing overwhelming material superiority, exhausted manpower, and an opponent who refused to make mistakes. Most scholars conclude that while Lee did about as well as anyone could have under the circumstances, no realistic path to escape existed after Five Forks.

Summary Comparison

Grant: One of the most perfectly executed pursuits in military history; ruthless, rapid, and almost flawless coordination. Frequently cited in military colleges as exemplary campaign design.

Lee: Conducted a skillful fighting retreat with an army on the verge of dissolution, but strategic misjudgments (especially the Amelia Court House delay) and inability to shake Sheridan sealed his fate.

In the end, the Appomattox Campaign is remembered less as a contest of equals than as the moment when Grant's strategic dominance and the North's material advantage finally forced even Lee's extraordinary army into capitulation.

Casualties

Due to the rapidity and fluidity of the campaign casualty statistics were not as well recorded as in most battles. But the result is overwhelming in Grant's favor: for a loss of about 5,000 men he captured Lee's army of 45,000. The war in Virginia was over.

Appomattox Campaign Casualty Comparison												
	Union						Confederate					
	Strength	Killed	Wounded	Captured/ Missing	Total	%	Strength	Killed	Wounded	Captured/ Missing	Total	%
Petersburg	114,335				3,396	3%	45,000				5,000	11%
Sayler's Creek	26,000				1,148	4%	18,500				7,700	42%
Appomattox Station	4,000				121	3%	3,000				1,000	33%
Appomattox Courthouse	63,285				164	0%	28,856				28,856	100%
Total					**5,369**	**5%**					**45,000**	**100%**

Rating Lee and Grant's Performance in the Appomattox Campaign

Grant's overall performance rates a perfect 5 or *Excellent*. He accomplished his dual strategic objectives of 1) capturing Lee's army and 2) capturing Richmond—the capital of the Confederacy. Tactically, Grant's ordering of a general advance after the Union victory at Five Forks and his unrelenting leadership and coordination of his subordinates to chase and overrun Lee's army leading to the surrender also rates a 5 or Excellent score. Grant's casualties were minor also generating a perfect 5 score of Excellent.

It is tough to judge Lee on the Appomattox Campaign as his army was already decimated and Grant's forces outnumbered him more than 2:1. Strategically, Lee failed to prevent Grant's breakout and he failed to lead his army to safety. Strategy score of 2 or Poor. Tactically, Lee was outfought throughout the campaign as he failed to gain any initiative other than retreat. Tactical score also Poor at 2. As a result Lee surrendered his Army of Northern Virginia leading to 100% casualties for a Casualty score of 1 or Bad. Overall score for Lee: 1.67 *Bad to Poor*.

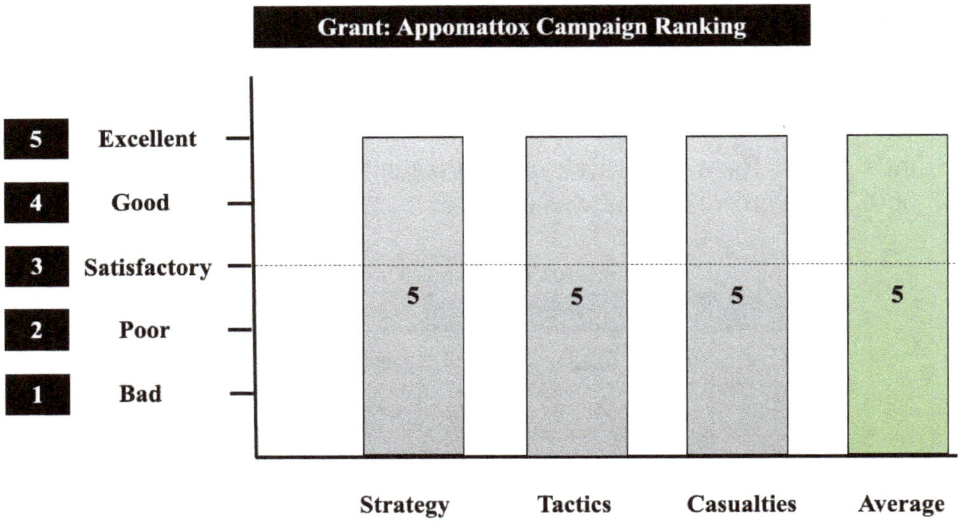
Grant: Appomattox Campaign Ranking
5 Excellent
4 Good
3 Satisfactory
2 Poor
1 Bad
5
5
5
5
Strategy
Tactics
Casualties
Average

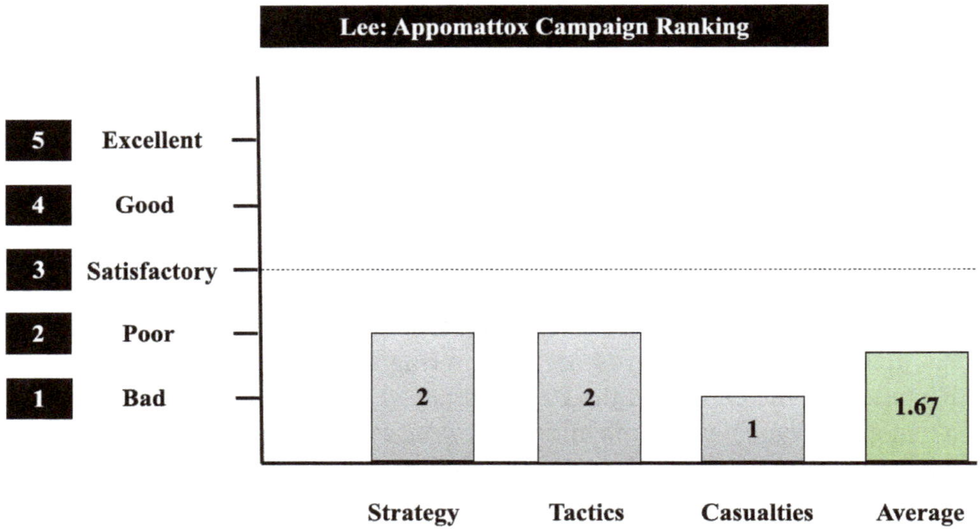
Lee: Appomattox Campaign Ranking
5 Excellent
4 Good
3 Satisfactory
2 Poor
1 Bad
2
2
1
1.67
Strategy
Tactics
Casualties
Average

David W. Manly

The Appomattox Campaign

Map by Hal Jespersen, www.cwmaps.com

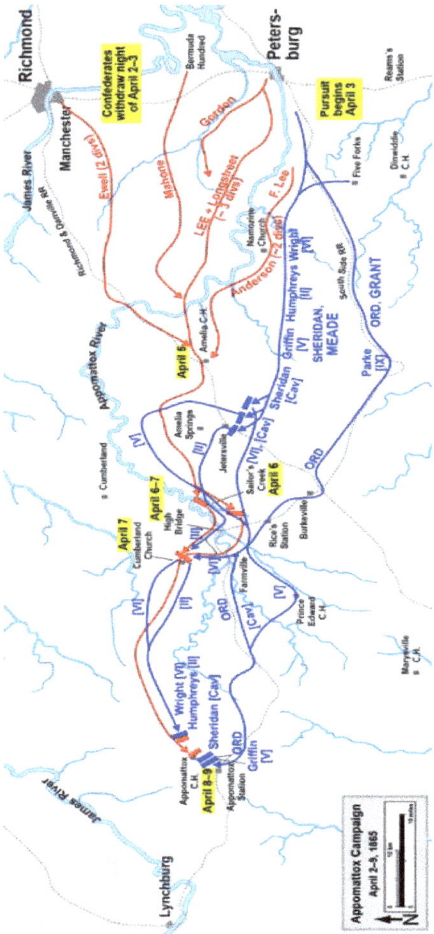

Conclusion

Now it is time to finally answer the critical question of this book: <u>who was the superior general in the United States Civil War—Lee or Grant.</u>

We shall address this question in two ways.

First, let's take a QUANTITATIVE look at the question based on the ratings data compiled in earlier chapters—a unique way of considering the question. After that we can consider more QUALITATIVE issues—similar to how most historians seek to answer the question.

Quantitative Review

We have reviewed the major battles and campaigns of General Lee and General Grant. We have graded their respective performances on the criteria of Strategy, Tactics, and Casualties for each major action. Now we can render a quantitative verdict: which general—Lee or Grant—was the superior general of the United States Civil War?

The table below summarizes the overall ratings given to each General in their respective battles/campaigns from the preceding pages. Here is what jumps out of the data:

- **<u>Overall our ratings have Grant exceeding Lee's rating</u>** at 4.05 vs. 3.33. Grant's performance in the war was *Good* averagely and Lee was *Satisfactory*.
- **<u>Grant's scores highest in Strategy</u>** (4.71) and lowest in Tactics and Casualties (3.71).
- **<u>Lee scores highest in Tactics</u>** (3.78) and lowest in Casualties (2.89).
- **<u>Interestingly, we rated Grant a perfect 5 (Excellent) in 3 battles but Lee achieved a perfect score in only two battles</u>**—Second Bull Run and Fredericksburg.
- **<u>Net, fairly close scores</u>** but "on the numbers" we must rank Grant's performance marginally higher than Lee's.

Grant and Lee Ratings Summary								
Battle/Campaign	Grant				Lee			
	Strategy	Tactics	Casualties	Average	Strategy	Tactics	Casualties	Average
Belmont	4	2	3	3				
Ft. Henry/Donelson	5	5	5	5				
Shiloh	4	3	2	3				
Seven Days					5	5	3	4.33
Second Bull Run					5	5	5	5
South Mountain					2	2	3	2.33
Antietam					2	4	2	2.67
Fredericksburg					5	5	5	5
Vicksburg	5	5	5	5				
Chancellorsville					5	5	4	4.67
Gettysburg					2	2	1	1.67
Chattanooga	5	4	4	4.33				
Overland Campaign	5	2	2	3	2	4	2	2.67
Appomattox Campaign	5	5	5	5	2	2	1	1.67
Average Score	**4.71**	**3.71**	**3.71**	**4.05**	**3.33**	**3.78**	**2.89**	**3.33**

The overall ratings are not the only quantitative data. Let's also dig deeper into casualty statistics for each General. After all, the key criticism of Grant after the war was that he was a "butcher" of his troops who had little strategic ability and who could not hold a candle to the more adept Lee. What do the numbers say about this?

Below is a comparison of casualties for all of General Grant's and General Lee's battles/campaigns—showing both losses sustained and losses generated on the enemy. The following can be discerned:

- In all his battles/campaigns, Grant suffered 103,314 total casualties yet inflicted 147,115 on the Confederates.
- Thus, Grant's own total casualties were 30% lower than those his armies inflicted on the enemy
- In all his battles/campaigns, Lee suffered 169,219 total casualties yet inflicted 180,656 on his Union Opponents.
- Thus, Lee's total casualties were only 6% lower than those his armies inflicted

Lee and Grant Casualty Comparisons								
	Union				Confederate			
	Killed	Wounded	Captured/ Missing	Total	Killed	Wounded	Captured/ Missing	Total
Lee's Battles Total	22,189	110,878	42,341	**180,656**	12,977	63,822	44,325	**169,219**
Grant's Battles Total	14,069	69,358	14,738	**103,314**	8,591	36,230	88,594	**147,115**

Lee Ratio: Confederate/Union: **94%** *Lee had 6% fewer losses than he inflicted*

Grant Ratio: Union/Confederate: **70%** *Grant had 30% fewer losses than he inflicted*

Net, on a macro level Grant was superior to Lee in his use of force to annhilate his enemy. This seems to indicate Grant is not the "butcher" post war Lost Cause historians labeled him as. If anything, Lee was more of a butcher than Grant. Certainly his failed charge at Gettysburg on Day 3 sacrificed the flower of his army as badly as anything Grant ever did.

To Lee's credit however, his armies suffered less killed (12,977) than the Union armies he faced (22,189) whereas Grant's armies suffered more killed than the Confederates they faced—14,069 to 8,591. Thus Grant's battles tended to be more ferocious and bloody perhaps than Lee's—adding some credence that perhaps Grant's tactics at time were "inefficient" at preserving his force.

Summary—quantitiatively we must give the edge to Grant as the superior general. His overall ratings were higher and on the macro level his armies caused more damage than they themselves suffered at a level superior to General Lee.

Quantitative Choice: U.S. Grant is the superior general of the Civil War.

Qualitative Considerations

Most historians who seek to answer the question of "who was best" study the question qualitatively—meaning their opinions rely less on data and more on an overall subjective picture based on many inputs. We can add qualitative analysis to our review to see how it dovetails with our quantitative conclusion. Specifically, we can compare our conclusion above with a balanced breakdown based on the most respected modern scholarship (Catton, McPherson, Bonekemper, Gallagher, Simpson, etc.). The consensus of these scholars is as follows:

Strategic Vision

Grant: Superior strategic vision. He was the only Union general who grasped from the beginning that the Civil War had to be a war against the entire Confederate economy and population, not just its armies. His 1864–65 coordinated offensive plan (simultaneous advances on multiple fronts) was the first truly national strategy either side attempted, and it crushed the Confederacy in 11 months.

<u>**Lee**</u>: Strong tactician and operational commander, but his strategic vision was narrow. He repeatedly advocated invading the North (1862 and 1863) to win a decisive battlefield victory or force peace negotiations — a political goal that never came close to succeeding and bled the Army of Northern Virginia white.

Winner: **Grant.** This is consistent with our quantitative review above.

Tactical/Operational Skill

<u>**Lee**</u>: Arguably the finest pure battlefield commander of the war. Chancellorsville (May 1863) is often called his "perfect battle." He consistently defeated larger Union armies in 1862–63 (Seven Days, Second Bull Run, Fredericksburg, Chancellorsville) with bold maneuvers and by taking enormous risks that usually paid off. But he was not infallible as we have noted at Gettysburg, Antietam and South Mountain.

<u>**Grant**</u>: Solid but not flashy. He rarely lost a battle after taking command in the West (Fort Donelson, Vicksburg, Chattanooga), but his Overland Campaign (1864) saw extremely high casualties because he maintained headlong attacks against entrenched opponents at battles like Spotsylvania and Cold Harbor and a second assault at Vicksburg.

Winner: **Lee** (but with a narrower margin than Lost Cause historians assign). Also consistent with our quantitative review above

Logistics, Administration, and Management of Resources

<u>**Grant**</u>: Outstanding. His Vicksburg Campaign (April–July 1863) is studied in military academies around the world for its logistical brilliance (living off the land, rapid river crossings, cutting his own supply line deliberately). He also managed the largest armies and theaters of the war without the breakdowns that plagued earlier Union commanders.

<u>**Lee**</u>: Decent, but he never had to manage anything like the scale Grant did, and he chronically overused his limited manpower (Pickett's Charge).

Winner: **Grant** decisively

Adaptability and Learning Curve

Grant: Learned from every setback (Belmont, Shiloh Day 1, Cold Harbor) and almost never repeated the same mistake.

Lee: Adapted less well in 1864–65. He continued seeking the knockout offensive blow even when he no longer had the manpower for it.

Winner: **Grant**

Results (the metric that ultimately matters most)

Grant: Won the war. Captured three Confederate armies (Fort Donelson, Vicksburg, Appomattox). Ended the rebellion in less than a year once Lincoln gave him supreme command.

Lee: Never took a major Union army captive after he assumed command of the Army of Northern Virginia. Lost the war and surrendered the Confederacy's most important field army.

Winner: **Grant** overwhelmingly

Modern Consensus Among Professional Historians

In virtually every poll of historians or military officers since the 1990s (including the 2013–2014 surveys by the U.S. Army War College and the American Political Science Association):

Grant usually ranks in the top 3–5 American generals of all time. Lee usually ranks 10–20, often behind Sherman, Eisenhower, Pershing, and sometimes even George Thomas or Winfield Scott.

Net: most serious students of the war today conclude Grant was the better overall general, and the gap has widened as the "Lost Cause" mythology about Lee has been dismantled over the last 40 years.

Final Word

The Civil War was won by the North because General Ulysses Grant designed a strategic plan to defeat the Confederacy at multiple points and ensured his armies in the field destroyed the Confederate armies in front of them—culminating in the March to the Sea by Sherman and Grant's victory at Appomattox. Strategically sound, and tactics that were efficient in the field.

On the other hand, the strategy and tactics of the Confederacy—as led by Lee—were not as in synch. The South had rebelled against the Union. It was now up to the Union to take action to compel the South into submission. All the South had to do was wait for the Union to attack/invade their territory and fight defensive battles on their own turf. Every time they did this—Bull Run, Chancellorsville—they badly mauled the Union armies.

So why feel compelled to invade the North? Why not just continue to repel the Union when it invaded? What if Lee had never invaded the North leading to a near disaster at Antietam and a full disaster at Gettysburg? Generating losses that were never fully made up. Can we not imagine that after several years of such a stalemate the Union would tire of the effort and call and end to it? And might the European powers finally recognize the Confederacy? We will never know.

But we can conclude based on Strategy...Tactics...Casualties along with qualitative analysis U.S. Grant was the superior general in the U.S. Civil War.

Epilogue

In closing, the reader may be interested in Lee and Grant's personal interactions, as well as their respective thoughts on each other.

Grant and Lee met personally four times during their lives. The first time they met was in Mexico during the Mexican/American War. The second, and most famous meeting, was at Appomattox where Lee surrendered to Grant. They also met briefly the following day when Grant rode out to meet Lee to discuss specific terms of the surrender ceremony. During this discussion Grant asked Lee to help arrange the surrender of Joe Johnston's Confederate Army as well but Lee demurred. Their last meeting was a courtesy call Lee made at the White House when Grant was President.

At Appomattox, Lee **a**rrived first around 1:00 p.m. accompanied only by his aide, Lt. Col. Charles Marshall. Grant arrived about 30 minutes later accompanied by several staff officers, including Ely S. Parker (a Seneca Indian and Grant's military secretary).

Grant and Lee at Appomattox

In his *Personal Memoirs* (Chapter 67), Grant provided a detailed, respectful account of his meeting with Lee at the McLean House in Appomattox Court House on April 9, 1865.

Grant described arriving mud-splattered and in a field uniform (due to a lingering migraine that eased upon receiving Lee's surrender note), while Lee appeared in a pristine full-dress uniform with an embroidered belt and his formal sword.

Grant wrote that he had served with Lee in Mexico but did not expect Lee (who was older and higher-ranking) to remember him. Grant initiated the topic to break the ice, noting: "I have always remembered your appearance, and I think I should have recognized you anywhere."

Lee responded that he also recalled the meeting but confessed he had often tried to picture Grant's features without success: "I have never been able to recall a single feature."

The two generals engaged in polite conversation initially about old army days, which Grant admitted nearly caused him to forget the purpose of the meeting. Lee eventually steered it to surrender terms.

Grant recounted feeling profound sadness rather than triumph:

"*I felt like anything rather than rejoicing at the downfall of a foe who had fought so long and valiantly, and had suffered so much for a cause, though that cause was, I believe, one of the worst for which a people ever fought*."

Grant emphasized the generous terms he offered—allowing Confederate officers to keep side arms and soldiers to retain horses/mules for farming—and Lee's gratitude for them, noting they helped ease soldiers' return home.

Overall, Grant portrayed Lee as dignified and composed in defeat, and the encounter as courteous, reflecting mutual respect between former comrades.

At the close of the surrender discussion both Grant and Lee exited the Mclean house together. Immediately the Union troops responded with celebratory firing of salutes with their weapons. Grant immediately stopped the celebration stating:

"*The Confederates were now our prisoners, and we did not want to exult over their downfall*."

Shortly after the surrender, Lee penned his final "Order" to the Army of Northern Virginia:

Headquarters, Army of Northern Virginia

April 10, 1865General Order No. 9

After four years of arduous service, marked by unsurpassed courage and fortitude, the Army of Northern Virginia has been compelled to yield to overwhelming numbers and resources.
I need not tell the survivors of so many hard-fought battles, who have remained steadfast to the last, that I have consented to this result from no distrust of them; but, feeling that valor and devotion could accomplish nothing that could compensate for the loss that must have attended the continuance of the contest, I have determined to avoid the useless sacrifice of those whose past services have endeared them to their countrymen. By the terms of the agreement, officers and men can return to their homes and remain there until exchanged. You will take with you the satisfaction that proceeds from the consciousness of duty faithfully performed; and I earnestly pray that a merciful God will extend to you His blessing and protection.
With an unceasing admiration of your constancy and devotion to your country, and a grateful remembrance of your kind and generous consideration of myself, I bid you an affectionate farewell. — R. E. Lee, General

The next day (April 10), Lee and Grant met briefly again on horseback to discuss details like parole passes and rations for Confederate troops (Grant provided 25,000 rations).

The formal surrender of the Confederate Army of Northern Virginia occurred on April 12, 1865, three days after the surrender. Neither Lee nor Grant attended this ceremony, which focused on the infantry formally laying down their arms and colors. Union Brigadier General Joshua Lawrence Chamberlain (a hero of Gettysburg) presided for the Union. Major General John B. Gordon represented the Confederates.

The Confederates marched in disciplined formation—regiment by regiment, with officers in place—carrying their arms and battle flags as if on parade. Many appeared ragged and emotional, some in tears, reflecting the weight of defeat after four years of war.

As the lead Confederate column approached, Chamberlain ordered his Union troops to "carry arms" (a marching salute, shifting rifles to the shoulder as a mark of respect). Gordon responded by ordering his men to return the honor, reportedly dropping his sword point in acknowledgment.

The Confederates then halted at designated points, faced the Union lines, and performed the ritual of surrender: bayonets fixed to muskets, rifles stacked in orderly piles, cartridge boxes unslung and placed on the stacks, and battle flags furled or folded (many soldiers kissed or tore pieces from their flags as keepsakes).

Officers retained their sidearms, and enlisted men who owned their horses or mules could keep them for farm work, per Grant's generous terms.

Approximately 26,000–28,000 Confederate soldiers participated, stacking around 27,000 stands of arms over the course of the day.

The atmosphere was somber and respectful rather than triumphant—no cheers or jeers from the Union side, only silence and mutual soldierly recognition.

Chamberlain later described it as "honor answering honor," a moment that helped begin national reconciliation.

This ceremony symbolized the effective end of major fighting in Virginia and set a model for subsequent Confederate surrenders.

The surrender ceremony at Appomattox

Lee and Grant's final meeting occurred shortly after Grant's inauguration as president in March, 1869. Lee, then president of Washington College (now

Washington and Lee University), was in Washington on railroad business and was invited by Grant to stop by the White House before he left the Capitol.

The encounter was brief, cordial but reserved, and focused on pleasantries rather than politics or the war. Accounts vary slightly due to conflicting newspaper reports, but reliable sources (including Lee's son and attendees) confirm no substantive issues were discussed. Grant is reported to have made a joke about the irony of them discussing post-war railroad business given how many railroads he and Lee destroyed in the war but the joke seemed to fall on deaf ears. They never met again as Lee died in 1870.

So what did Lee and Grant thing of each other?

In his *Personal Memoirs* Grant offered a measured, somewhat understated assessment of Robert E. Lee as a military commander. Grant sought to counter the widespread and growing mythologizing of Lee as an invincible or superhuman general. One of the most direct passages comes early in Volume 2, where Grant reflects on perceptions of opposing commanders:

"The natural disposition of most people is to clothe a commander of a large army whom they do not know, with almost superhuman abilities. A large part of the National army, for instance, and most of the press of the country, clothed General Lee with just such qualities, but I had known him personally, and knew that he was mortal; and it was just as well that I felt this."

This sets the tone: Grant viewed Lee as capable but human and fallible, not the demigod many portrayed him as.

Later, in discussing the Overland Campaign of 1864 he compared Lee favorably to other Confederate generals but ranked him below Joseph E. Johnston in terms of causing him personal anxiety:

"I never ranked Lee as highly as some others of the army—that is to say, I never had as much anxiety when he was in my front as when Joe Johnston was in my front. Lee was a good man, who had everything in his favor."

Grant acknowledged Lee's advantages: fighting on defensive terrain he knew intimately, with an experienced army, and receiving reinforcements roughly equal to Union gains (minus expirations of service). He implied that

Lee's successes were amplified by these factors and the aura built around him, rather than pure tactical superiority.

Grant also noted Lee's bold offensives (e.g., invasions of the North), but suggested they were risky and ultimately unsustainable given the Confederacy's resource disadvantages. Overall, Grant portrayed Lee as a strong defensive commander who inflicted heavy casualties but could not overcome the Union's strategic advantages in manpower and logistics.

Grant's tone is respectful—he describes Lee as dignified during their meetings at Appomattox and expresses sorrow at the Confederacy's defeat—but he consistently demythologizes him, emphasizing that Lee's reputation was inflated and that Grant's persistent pressure (despite high costs) proved decisive.

These views reflect Grant's broader philosophy: war was about sustained, coordinated pressure across theaters, not individual brilliance in isolated battles. Modern historians often note that Grant's assessment was partly defensive (countering post-war "Lost Cause" glorification of Lee), but it aligns with his pragmatic, no-nonsense style throughout the memoirs.

And what did Lee think of Grant?

Unfortunately, Lee did not publish any memoirs, or even articles about his Civil War activities. It almost seems he wanted to move on and not relive a tragic period of his life. So we have no reputable <u>direct</u> knowledge of what Lee thought of Grant as an opponent.

However there is some tangential information that Lee did not consider Grant as his toughest Union adversary. According to Robert E. Lee Jr.'s (General Lee's son) *Recollections and Letters of General Robert E. Lee*, published in 1904 General Lee is reported to have considered George McClellan to be the best Union general he faced. Specifically the book states that after the war ended, a son of one of Lee's cousins asked the former general "which of the Federal generals he considered the greatest," to which Lee "answered most emphatically '*McClellan by all odds*.'"

*****

It is interesting that both Grant and Lee felt a need to downplay each other—when it is so clear over a hundred and fifty years later that the two of them stand alone as the giants of the American Civil War.

Last Known Photograph of Grant Days Before His Death from Cancer

Last Photograph of Lee: August 31, 1870

Bibliography/Suggested Reading

Biographies of Ulysses S. Grant

Here are the most highly recommended modern biographies, based on reviews from historians, presidential biography experts, and Civil War scholars:

- **Grant** by Ron Chernow (2017) — Widely regarded as the definitive modern biography. Comprehensive, well-researched, and readable; it rehabilitates Grant's reputation as a general and president while addressing his flaws like alcoholism.
- **Grant** by Jean Edward Smith (2001) — Often tied with Chernow's as the best; balanced, insightful, and a Pulitzer finalist.
- **Personal Memoirs of Ulysses S. Grant** by Ulysses S. Grant (1885–1886) — Essential primary source and one of the greatest military memoirs ever written. Focuses on his life up to the Civil War's end; remarkably candid and eloquent.
- **American Ulysses: A Life of Ulysses S. Grant** by Ronald C. White (2016) — Excellent single-volume overview, emphasizing Grant's character and post-war life.
- **Grant** by William S. McFeely (1981) — Pulitzer Prize winner; more critical of Grant but still highly respected for its depth.
- **The Man Who Saved the Union: Ulysses S. Grant in War and Peace** by H.W. Brands (2013)--A biography of Ulysses S. Grant, focusing on his role in the later Civil War and presidency.
- **Grant at Vicksburg: The General and the Siege** by Michael Ballard (2013)
- **Ulysses S. Grant, Triumph Over Adversity, 1822-**1865, by Brooks D. Simpson (2000)
- **U.S. Grant: The Making of a General by** Michael Ballard (2005)
- **General Grant and the Verdict of History: Memoir, Memory, and the Civil War** by Frank P. Varney (2023)

Biographies of Robert E. Lee

Top recommendations emphasize balanced, modern perspectives over older hagiographic works:

- **Robert E. Lee: A Biography** by Emory M. Thomas (1995)—Considered the best single-volume biography; balanced, humanizes Lee without mythologizing, and draws on extensive sources.
- **Robert E. Lee: A Life** by Allen C. Guelzo (2021) –Nuanced and recent; explores Lee's complexities, including his decisions on secession and slavery, with strong contextual analysis.
- **R.E. Lee: A Biography** by Douglas Southall Freeman (1934–1935, 4 volumes; often read in abridged form)–Classic and detailed Pulitzer winner, but pro-Lee and dated—best for depth if you want the traditional view.
- **Reading the Man: A Portrait of Robert E. Lee Through His Private Letters** by Elizabeth Brown Pryor (2007)—Unique insight into Lee's personality via his own words; reveals flaws alongside strengths.
- **Lee Considered: General Robert E. Lee and Civil War History** by Alan Nolan (1991)

Books Comparing or Covering Both Grant and Lee--for direct comparisons of their leadership, strategies, and legacies:

- **Crucible of Command: Ulysses S. Grant and Robert E. Lee—The War They Fought, the Peace They Forged** by William C. Davis (2014) — Excellent dual biography contrasting their lives, military styles, and post-war roles.
- **Grant and Lee: Victorious American and Vanquished Virginian** by Edward H. Bonekemper III (2007) — Analytical comparison of their generalship, arguing Grant's approach was more effective for victory.
- **Grant and Lee**: A Study in Personality and Generalship by J.F.C. Fuller (1933)
- **A Place Called Appomattox** by William Marvel (2000)
- **An English View of the Civil War** by Lord Garnet Wolseley (1889)

- **And Keep Moving On: The Virginia Campaign May-June, 1864** by Mark Grimsley (2002)
- **Battle Cry of Freedom: The Civil War Era** by James McPherson (2003)
- **Bloody Roads South: The Wilderness to Cold Harbor, May-June 1865** by Noah Andre Trudeau (1989)
- **Civil War Supply and Strategy: Feeding Men and Moving Armies** by William Marvel (2024)
- **Cold Harbor: Grant and Lee**, May 26-June 3, 1864 by Gordon Rhea (2002)
- **Confederate Tide Rising: Robert E. Lee and the Making of Southern Strategy 1861:1862** by Joseph L. Harsh (1998)
- **Davis and Lee at War** by Stephen Woodworth (1995)
- **Gettysburg** by Stephen Sears (2003)
- **Landscape Turned Red: The Battle of Antietam** by Stephen Sears (1983)
- **Nothing but Victory: The Army of the Tennessee 1861-1865** by Stephen Woodworth (2005)
- **Race to the Potomac: Lee and Meade after Gettysburg, July 3-14, 1863** by Bradley M. Gottfried and Linda I. Gottfried (2024)
- **Recollections and Letters of General Robert E. Lee** by Robert E. Lee, Jr. (1904)
- **Six Armies in Tennessee: The Chickamauga and Chattanooga Campaigns** by Stephen Woodworth (1998)
- **Taken at the Flood: Robert E. Lee and Confederate Strategy in the Maryland Campaign of 1862** by Joseph L. Harsh (1999)
- **The American Civil War** by Gary Gallagher (2000)
- **The Battle of the Wilderness**, May 5-6, 1864 by Gordon Rhea (1994)
- **The Campaign for Vicksburg** by Edwin Bearss (1991)
- **The Civil War in the West** by Earl J. Hess (2012)
- **The Enduring Civil War** by Gary Gallagher (2020)
- **The Gettysburg Campaign: A Study in Command** by Edwin Codington (1997)
- **This Great Struggle: America's Civil War** by Stephen Woodworth (2011)

David W. Manly

About the Author

David (Dave) Manly is a retired business executive who passionately studies the Civil War as a key hobby after discovering both sides of his family participated in the War.

Prior to his retirement in 2015, Dave was a senior executive at several consumer companies with well-known brands. Dave was one of the key executives that took *Keurig* from start-up to the number one selling coffee maker in North America greatly improving the daily coffee ritual for millions. Dave helped *LoJack* (stolen vehicle recovery) become a household name and gold standard in auto security. At the *Boston Whaler* boat company, Dave brought innovative consumer marketing to the boating industry. Dave started his career at Procter & Gamble where he served as Brand Manager of *Pringle's Potato Chips*, and *Crisco Oil.*

Dave holds a B.A. from DePauw University (1977), and a Master of Science in Marketing (MSM) from the Krannert (now Daniels) School of Business at Purdue University (1979). Dave and his wife Marcia reside south of Boston in Duxbury, MA and Martha's Vineyard close to their three children and grand-children.

www.ingramcontent.com/pod-product-compliance
Lightning Source LLC
LaVergne TN
LVHW082348200926
848481LV00012B/3494
* 9 7 9 8 9 9 5 6 3 2 8 1 8 *